PENGUIN BOOKS

CIVILIZING THE MACHINE

John F. Kasson was born in Muncie, Indiana, in 1944. He earned his bachelor's degree in history at Harvard in 1966, and in 1971 he received his doctorate in American studies at Yale. Currently, he teaches in the Department of History at the University of North Carolina at Chapel Hill. *Civilizing the Machine* is his first book.

John F. Kasson

CIVILIZING
THE MACHINE

Technology and Republican
Values in America
1776-1900

Penguin Books

CIVILIZING THE MACHINE

PENGUIN BOOKS
Published by the Penguin Group
Viking Penguin Inc., 40 West 23rd Street, New York, New York 10010, U.S.A.
Penguin Books Ltd, 27 Wrights Lane, London W8 5TZ, England
Penguin Books Australia Ltd, Ringwood, Victoria, Australia
Penguin Books Canada Ltd, 2801 John Street,
Markham, Ontario, Canada L3R 1B4
Penguin Books (N.Z.) Ltd, 182–190 Wairau Road,
Auckland 10, New Zealand

Penguin Books Ltd, Registered Offices:
Harmondsworth, Middlesex, England

First published in the United States of America
by Grossman Publishers 1976
Published in Penguin Books 1977

9 11 13 15 14 12 10 8

LIBRARY OF CONGRESS CATALOGING IN PUBLICATION DATA
Kasson, John F.   1944-
Civilizing the machine
1. Technology—Social aspects—United States—
History.   2. United States—Civilization—19th
century.   I. Title.
[T14.5.K37 1977]   301.24'3'0973   77–394
ISBN 0 14 00.4415 9

Printed in the United States of America
Set in Janson

Acknowledgments

Harvard University Press: From *The Journals and
Miscellaneous Notebooks of Ralph Waldo Emerson*,
edited by William Gilman et al. Copyright © 1964, 1966, 1971, 1973
by the President and Fellows of Harvard College.
Houghton Mifflin Company: From *The Journals of Ralph Waldo
Emerson*, edited by Edward Waldo Emerson
and Waldo Emerson Forbes.
Copyright 1912, 1913 by Edward Waldo Emerson.

For Joy

Preface

The Gallery of Machines at the French World's Fair of 1900 inspired Henry Adams to contemplate the sweep of history from the Virgin, the figure of medieval energy, to the dynamo, the symbol of the modern age. The present historical essay is more modest in scope than Adams's, but it too was stimulated by a gallery of machines, a visit in the fall of 1968 to an exhibition at the Museum of Modern Art: "The Machine—As Seen at the End of the Machine Age." The magnificent variety of artists' responses to machine technology prompted me to examine the history of America's own cultural response to technology. The farther my investigation progressed, however, the more I discovered that this response could not be understood in isolation; that the expectations Americans have historically brought to their technology are profoundly rooted in their understanding of the entire republican experiment.

New appreciation of the importance of republicanism has been one of the most interesting and fruitful developments in American historical writing in recent years. A generation ago historians generally viewed the origins of American republicanism in narrow political or intellectual terms. In the past two decades, however, studies by a number of historians, particularly Bernard Bailyn and Gordon S. Wood, have combined to provide a very different understanding of republicanism, one which views it as a dynamic *ideology* concerned with the maintenance of liberty, order, and virtue and assuming social and cultural dimensions such as to make it central to American civilization. This reformulation has not only changed the whole terms of our understanding of the Revolutionary period; it also carries important implications for later American history which are only now beginning to be realized.

The problem upon which this study concentrates—addressed on various levels from the Revolution throughout the nineteenth century—is the meaning of technology for a republican civilization. The precise terms of America's republican ideology were never firmly fixed. Quite the contrary, in its very fluidity, republicanism formed the subject for intense debate from the moment of the Revolution onward, ranging between conservative warnings of an excess of democracy and egalitarian complaints of its inadequate fulfillment. The rapid development of machine technology and the process of industrialization as a whole altered the context of this discussion and fundamentally tested the country's republican commitment on a number of levels. As Americans reflected upon the proper place of technology in a republic, they were compelled to articulate the kind of civilization they desired for the nation. They had to re-examine their conceptions of the entire social order and determine how best to maintain social cohesion and purity. Concern for the social consequences of industrialization sparked renewed consideration of the degree of social opportunity and the meaning of egalitarian principles in a technological society. Technology raised equally vital questions for the imaginative and cultural life of the nation. New machinery and modes of communication enormously expanded the range of human perceptions, but they also threatened to dull the individual conscience and creative spirit. As technology dramatically reshaped the physical environment, it also transformed Americans' very notions of beauty and raised critical issues of the proper art form for a republic. By the closing decades of the nineteenth century, these and related concerns swelled to a climax as the future of republicanism in a technological age appeared hedged at once by millennial hopes and bitter doubts.

In order to examine the impact of technology upon republican values in some detail, I have chosen to focus upon a number of specific topics: the debate over the introduction of domestic manufactures in the Revolutionary period; the divided response to the model factory town of Lowell, Massachusetts, in the second quarter of the nineteenth century; Ralph Waldo Emerson's internal dialogue on the impact of technology upon the American imagination; aesthetic responses to and claims for machinery; and utopian and dystopian novels of the 1880s and '90s. Each of these subjects powerfully illuminates a significant aspect of the impact of technology on the republic. Taken together, they assume an even larger importance. They demonstrate a persistent concern with the meaning of technology for American civilization during a major period of the nation's history and reveal the terms of that discussion. They show both the way in which American technological development and republican values profoundly shaped one another and the difficulty—ultimately leading to failure—of achieving a technological society consonant with republican ideals.

I am indebted to a number of individuals and institutions for assistance in the writing of this book. First, to the American Studies Program at Yale University, which provided the stimulus to begin this project; next, to the National Endowment for the Humanities for a fellowship to complete it. For assistance in my research and in gathering illustrations, I would like to acknowledge the staffs of the Boston Public Library; the Duke University libraries; the Harvard University libraries; the Henry Ford Museum; the Library of Congress; the Manchester Public Libraries; the Mark Twain Memorial; the Massachusetts Historical Society; the Merrimack Valley Textile Museum; the Metropolitan Museum of Art; the National Gallery of Art; the National Museum of History and Technology; the New-York Historical Society; the New York Public Library; the Putnam County Historical Society; the Science Museum, London; the libraries of the University of North Carolina at Chapel Hill; and the Yale University Art Gallery and libraries. I am also grateful for the support of a number of friends, colleagues, teachers, and students. I wish particularly to thank Henry Abelove, Peter Filene, David D. Hall, Neil Harris, A. N. Kaul, Donald G. Mathews, Daniel Okrent, Jules Prown, Daniel Rodgers, Neil Shister, and Alan Trachtenberg for their criticism and encouragement at various stages of composition. While I have attempted to indicate specific intellectual debts in the notes at the back of the book, I would like to declare a special sense of obligation and inspiration to the

work of Bernard Bailyn, Leo Marx, Perry Miller, and Henry Nash
Smith. By my dedication I have gestured to that which I cannot
hope to express: my deep gratitude to my wife, Joy S. Kasson,
for her abundant enthusiasm, stimulating criticism, and sustaining
faith.

Contents

Illustrations

1

The Emergence of Republican Technology

I N THE late eighteenth century America began not one revolution but two. The War of American Independence coincided with the advent of American industrialization, and these dual transformations ultimately conjoined in a way that has shaped the character of much of our history. The ideological links between technology and republicanism, however, were only gradually forged. For the new nation, the meaning of both the Revolution and modern machinery was uncertain and the relationship between the two problematic. Dispute over the implications of technology for the new nation involved critical issues of American destiny. Influential citizens argued whether introduction of domestic manufactures would ensure America's political independence, economic and social stability, and moral purity or subvert them; whether technology would help to integrate the country into a cohesive nationality or prove a divisive agent. These and related questions were anxiously discussed throughout the entire Revolutionary period, from the Stamp Act crisis through the ratification of the Constitution, by men highly conscious of the precariousness of the republican venture. Yet within a half century such questioning was pursued only by a relative few. For the dominant voices in public discussion, doubt was unthinkable; they hailed the union of technology and republicanism and celebrated their fulfillment in an ever more prosperous and progressive nation. The transition between these two positions, whereby technology came to be regarded as essential to American democratic civilization, was of fundamental importance to the nation's history.

Examination of this dual revolution of politics and technology must begin with the ideological background of the American Revolution. Ideology alone, of course, did not cause either of these great transformations, but it shaped the understanding that formed the basis for action. It constituted the configuration of assumptions, values, and beliefs by which Americans sought to interpret and transform their political and social relationships. The center of

American Revolutionary ideology was the concept of republican-ism.[1] For the Revolutionary generation, republicanism was not an intellectual superstructure, not an external, artfully contrived weapon of propaganda, despite the manner in which later historians would perceive it. Nor was it simply an invention of the moment. The roots of republican ideology extended deep into English politics and the English libertarian tradition, Puritanism, Enlight-enment rationalism, ancient history and philosophy, and common law, and these roots were strengthened rather than severed by be-ing transplanted into the fertile soil of America. The notion of republicanism began with a conception of the relationships among power, liberty, and virtue. The balance among these elements, Americans' reading and experience taught them, remained deli-cate and uneasy at best. Power, as they conceived it, whether wielded by an executive or by the people, was essentially aggres-sive, forever in danger of menacing its natural prey, liberty or right. To safeguard the boundaries between the two stood the fundamental principles and protections, the "constitution," of government. Yet this entire equilibrium depended upon the strictest rectitude both within government and among the people at large. To the eighteenth-century mind republicanism denoted a political and moral condition of rare purity, one that had never been successfully sustained by any major nation. It demanded extraordinary social restraint, what the age called "public virtue," by which each individual would repress his personal desires for the greater good of the whole. Public virtue, in turn, flowed from men's private virtues, so that each individual vice represented a potential threat to the republican order. Republicanism, like Puri-tanism before it, preached the importance of social service, in-dustry, frugality, and restraint. Their opposing vices—selfishness, idleness, luxury, and licentiousness—were inimical to the public good, and if left unchecked, would lead to disorder, corruption, and, ultimately, tyranny. The foundation of a just republic con-sisted of a virtuous and harmonious society, whose members were bound together by mutual responsibility.[2]

These ideas, derived from a vigorous dissenting tradition within English politics, were transformed into a dynamic revolu-tionary ideology in the turbulent years after the French and

Indian War. Keenly sensitized to abuses of power, obsessed with the decline of previous republics, and aroused to defend their liberties and virtue against any aggression, the colonists perceived the increasing severity of British colonial rule as not simply an unjust and onerous policy, but a malevolent conspiracy. In every forum and by every mode of expression available—in pamphlets, newspapers, broadsides, almanacs, sermons, speeches, treatises, letters, and diaries—men considered the meaning of their social and political experience, their relations with England, and what they came to see as their special destiny in the world. Republican ideology provided a comprehensive explanation, a coherent instrument of analysis, and a path of resistance to the disturbing flow of events of the Revolutionary era. It gathered scattered events and grievances into a unified and comprehensible pattern of oppression. It crystallized and mobilized discontent, articulated and symbolized political and social ideals, and offered direction toward the establishment of a new political and cultural order. Republicanism, in short, provided the catalyst and shaped the character of the American Revolution.[3]

Republican ideology not only propelled the breach with England; it transformed America as well. Stimulated by the need to examine and articulate their disputes with Britain, forced to explore implications in their thought and experience, Americans moved with astonishing speed in the Revolutionary era toward positions none would have anticipated only a few years before. Concepts of political representation and consent, of constitutions, rights, and sovereignty were all dramatically recast, and attitudes toward established religion, chattel slavery, and social hierarchy significantly modified. Republican ideology led finally beyond politics to a major coalescence and reorientation of American culture. The Revolutionary spirit charged virtually every aspect of life. It lifted Americans to an almost utopian strain. "Novus Ordo Seclorum" trumpeted the Great Seal: a new order of the ages. Americans believed themselves not only engaged in a momentous political experiment, but embarked on a new civilization.[4]

In such a momentous, indeed millennial, undertaking, virtually every decision Americans faced was invested with crucial significance. If preservation of public virtue was essential to good government and the social order, then beneath every temptation lurked possible treason, beneath every deviation potential disaster. Despite a common ideological commitment, the nation had no blueprint as to how to proceed. Apprehension thus mixed with exultation, and the result of the republican consensus was not harmonious agreement but a highly charged, anxiety-ridden political climate in which men unceasingly called for moral reformation while charging their opponents with subversion.[5]

It is within such a context that one must view the emergence of republican technology. The questions of the introduction of domestic manufactures and the role that labor-saving machines might play in American life were considered not as isolated economic issues but as matters affecting the entire character of society. No doubt profit motives existed, but would-be manufacturers had to make cogent arguments which addressed broader ideological concerns. In addition to asking, "How much will it pay?" they had to consider as well, "How will it advance the cause of republicanism?" The question was not rhetorical—not at this time at least. That stage of the discussion would be reached only by the Revolutionary generation's grandchildren.

The history of the emergence of republican technology cannot be fully grasped unless one first understands the extent to which Americans on the eve of the Revolution conceived of their identity in terms of agriculture. The country stretched before them, an apparently inexhaustible wilderness to be subdued and cultivated. One could begin to smell pine trees as far as 180 nautical miles from land, and the forests extended far beyond the Appalachian frontier until they trailed off in legend. Such a fertile and spacious setting spurred extraordinary growth. From a population of approximately 250,000 in 1700, the colonies had swelled to 1,170,000 by 1750, 2,148,000 by 1770, and by the end of the century the United States would boast of more than five million people. Roughly 80 per cent of the population worked on farms, and many others depended directly upon the farming economy for their living. Colonial Americans, of course, shared quite unequally in the prosperity of the New World. Yet despite the great extremes of conditions separating aristocrats, indentured servants, and slaves, the colonies had achieved to an unprecedented degree a middle-class society based on their rural economy.[6]

In such a society farming constituted a source of cultural value and a sign of virtue, a moral as well as an economic condition. Praise of husbandry, of course, has an ancient literary heritage extending back to Hesiod. In the eighteenth century educated Americans drew liberally upon this tradition for guidance and precedent. The literature of Augustan and Imperial Rome upon which they concentrated paid tribute to farming as the foundation of liberty and virtue, so that they could join Cicero in declaring, "Of all the occupations by which gain is secured, none is better than agriculture, none more profitable, none more delightful, none more becoming a freeman."[7]

Under the pressure of ideas and events in the years after 1765 this image of agrarian America was intensified and transformed into a revolutionary symbol of republican virtue. Revolutionary spokesmen placed special emphasis upon what might be called the ecology of liberty. Americans, they maintained, stood in a particularly strategic position as the defenders of freedom against corruption and tyranny because they were uniquely favored by nature and history in the abundance of land and tradition of independence essential for a free people. The independent yeoman was elevated as the symbolic hero of the American struggle and the farmer became a favorite persona of revolutionary literature. In such a role J. Hector St. John de Crèvecoeur described the way in which a European immigrant was transformed by his new environment: A man comes to America, hires himself out to a farmer, works well and is treated with respect. After he has established himself, he buys land and begins a farm of his own. "What an epocha in this man's life! He has become a freeholder, from perhaps a German boor.... From nothing to start into being; from a servant to the rank of a master; from being the slave of some despotic prince, to become a free man, invested with lands to which every municipal blessing is annexed! What a change indeed! It is in consequence of that change that he becomes an American."[8] It was a normative as well as a descriptive definition of an American that Crèvecoeur proposed. A republican society was a society of freeholders, and praise of husbandry amounted to a national faith.

But if agriculture assumed a position of special importance in the republican enterprise, so did technology. The word "technology," of course, did not acquire its current meaning until the nineteenth century. In eighteenth-century usage "technology" denoted a treatise on an art or the scientific study of the practical or industrial arts, but not the practical arts collectively.[9] Indeed, men made little or no distinction in this period between theoretical science and mechanical ingenuity, contentedly grouping both under the rubric of "useful knowledge." When, for example, in 1743, Benjamin Franklin called for the establishment of the colonies' first learned society, he titled his paper *A Proposal for Promoting Useful Knowledge among the British Plantations in America*. Intended to pursue "all philosophical Experiments that let light into the Nature of Things, tend to increase the Power of Man over Matter, and multiply the Conveniencies or Pleasures of Life," the proposed society would range from botany and mathematics to labor-saving inventions and manufactures. For all knowledge was deemed related and advantageous, and even those whose investigations did not aim at specific, practical ends remained confident that whatever truths they uncovered would ultimately prove useful. Observing the first balloon ascent in Paris, Franklin heard a scoffer ask, "What good is it?" He spoke for a generation of scientists in his retort, "What good is a newly born infant?" He and his American contemporaries shared the Enlightenment assumption that as knowledge increased, so inevitably would its practical applications and rewards, and thus they took steps to promote the two together.[10]

With the struggle for American independence the question of utility took on a new dimension: technology emerged as not merely the agent of material progress and prosperity but the defender of liberty and instrument of republican virtue. This process began in the decade of colonial agitation between passage of the Sugar and Stamp Acts in 1764–65 and the outbreak of open revolution. During this period Parliament's repeated efforts to tax Americans were opposed through various boycotts against British goods. Such movements were designed, as Edmund S. Morgan has observed, as a mode at once of resistance and of reaffirmation. Colonists began to see a threat to their liberty not only in the tax laws themselves, but also in the items which bore the tax and even in British goods in general. Increasingly, they voiced to one another the fear that England plotted to undermine their love of liberty by a double strategy: the practices of an oppressive government (as exemplified in the taxes) and the infection and cultivation among colonists of the deadly taste of luxury (through consumption of British wares). Nonimportation and nonconsumption thus emerged as an essential program in the defense of republican virtue, organized and implemented in communities throughout the colonies, and so too did domestic manufactures. Instead of buying "foreign trifles," colonists vowed to encourage industry, frugality, and independence through the production and consumption of native goods. "If we mean still to be free," they told one another, "let us unanimously lay aside foreign superfluities, and encourage our own manufacture. SAVE YOUR MONEY AND YOU WILL SAVE YOUR COUNTRY."[11] Leading citizens of various communities discarded British silks and broadcloths in favor of homespun, and a "spinning craze" swept through women's circles throughout the colonies. Towns and colonial legislatures offered bounties, tax exemptions, and other inducements to stimulate domestic production. The Boston Society for Encouraging Industry and Employing the Poor and the New York Society for the Promotion of Arts, Agriculture, and Oeconomy organized spinning schools and established America's first rudimentary textile factories. Resistance to English authority triggered the greatest flurry of domestic manufacturing America had ever seen.[12]

These efforts were capped in 1775 by the founding of the United Company of Philadelphia for Promoting American Manufactures, which employed nearly five hundred people in the making of cotton, linen, and woolens. On March 16 the company's president, the distinguished scientist and physician Benjamin Rush, delivered a speech to the subscribers which offered the most extended justification of republican manufactures of the time. The advantages of promoting domestic manufactures, in Rush's eyes, were considerable. By such a policy, America would increase her wealth, lessen consumption of European luxuries, attract skilled labor from Europe, provide useful employment for the poor, and, most important, "erect an additional barrier against the encroachments of tyranny. A people who are *entirely* dependant upon foreigners for food or clothes," Rush warned, "must always be subject to them." Such dependency easily became slavery, wherein Americans would lose all principle of virtue and all power to resist English tyranny. Native manufactures thus provided an essential defense of freedom. In the face of such advantages, Rush argued, objections were chimerical. Promotion of manufactures would not lessen the country's commitment to agriculture, since less labor would be needed for importation of goods, and manufacturing would enlist the talents of men not attracted to agriculture. Moreover, Rush calculated that, properly conducted, two-thirds of the manufacturing labor force would be women and children. At the same time, he assured his audience that the diseases and misery associated with British manufacturing stemmed from "unwholesome diet, damp houses, and other bad accommodations" rather than from manufacturing in itself. Such extrinsic evils might be avoided in America, he declared, adumbrating what would become a major theme in the discussion of early nineteenth-century American factory towns. As a clinching argument, Rush proudly gestured to the anticipated impact of the United Company's newly imported spinning jenny, the first recorded jenny in America. Such machines, he was confident, would enable America to manufacture woolens and linens of superior quality at a cost comparable to English imports. Speaking a month before the battles of Lexington and Concord, Rush in his conclusion did not absolutely dismiss the possibilities of reconciliation, but in effect he called for the use of new technology and the promotion of manufactures as the essential foundation of American independence.[13]

Such efforts at large-scale manufacturing were a symbol and prophecy of the unfolding revolution in textile technology and industrial organization then taking place in Great Britain and soon to transform American production. For the time being, however, the great bulk of American manufacturing continued to take place in home and workshop rather than factory. (Indeed, in this context, "domestic" in a double sense, manufacturing was perhaps more easily appropriated into the rhetoric of republicanism.) Altogether, the nonimportation movement and its accompanying encouragement of domestic manufactures had important implications for America's future. Manufacturing had emerged from the shadows of mercantile disfavor to a position of new centrality in the American economy, and the argument that America must remain forever dependent upon the Old World for its goods lost much of its earlier conviction. Equally important, domestic manufactures had assumed new significance in the swelling ideology of American republicanism. The enhanced status of manufacturing as a fit companion of agriculture among republican occupations and as an essential agent in the defense of American virtue from external corruption would have critical repercussions in later American economic and social development.

The emerging alliance between technology and republicanism assumed another stage with the outbreak of war in 1775. Domestic manufactures were no longer a gesture of protest but a military necessity, despite the invaluable assistance of foreign goods. In March 1776 the Continental Congress passed a resolution drawn up by John Adams urging every colony to establish "a society for the improvement of agriculture, arts, manufactures, and commerce, and to maintain a correspondence between such societies, that the rich and numerous natural advantages of this country, for supporting its inhabitants, may not be neglected."[14] In response to this and similar calls on state and local levels, earlier efforts to stimulate home productions were extended and generalized. At the same time the new nation launched upon the task of devising and manufacturing armaments for the revolutionary struggle. America already possessed an important iron industry, which accounted in 1775 for one-seventh of the world's total output of crude iron. Steel production, less developed, made rapid strides during this time; by the Treaty of Versailles twice as many steel plants stood as had existed in 1750. Other industries similarly mushroomed to meet the challenge of war. Once hostilities began, for example, colonists from the Continental Congress on down exerted every possible effort to facilitate production of saltpeter (potassium nitrate), a vital ingredient in gunpowder which America had not hitherto produced. Although during the first half of the war the colonies continued to depend largely upon foreign gunpowder, domestic production steadily increased. By 1786 Pennsylvania alone had established twenty-one powder mills with an annual capacity of 625 tons. All the while a host of foundries sprang up throughout the colonies to manufacture arms and ammunition, many stimulated by government aid and contracts, such as those at Springfield, East Bridgewater, and Easton, Massachusetts; Trenton, New Jersey; Lancaster, Pennsylvania; and Principio, Maryland. The Revolution demanded that Americans defend their republican virtue by manufactures as well as agriculture, and thus the nation beat its plowshares into swords.[15]

In devising weapons and military strategy, American forces were handicapped by the lack of a trained corps of military engineers. George Washington bewailed this deficiency in a letter to the Continental Congress within a week after he had assumed his position as commander-in-chief of the army in 1775; and one of Franklin's first errands upon arriving as minister to France in December 1776 was to persuade the French minister of war to lend the services of four French engineers.[16] In lieu of a professional American engineering corps, the revolutionaries had to rely instead upon the efforts and proposals, often independent and uncoordinated, of native scientists and inventors. The source of these various suggestions, both practical and hypothetical, and the manner in which they were forthcoming, offers striking insight into the Revolutionary participants' sense of the relationship between technology and republicanism.

In a society which could not afford the luxury of specialization or conceive of knowledge as narrowly compartmentalized, proposed innovations and improvements of weapons flowed from men of considerable achievement in diverse fields. Among those who threw themselves most completely into Revolutionary service was the mathematician, astronomer, and instrument-maker David Rittenhouse, America's most distinguished scientist after Franklin. Suspending his scientific research by which he had constructed the finest orrery, or mechanical planetarium, to date and made important observations of Venus, he turned his attention to military and political matters pressed upon him by contemporaries who believed in the utility of science and the continuity of knowledge.

Rittenhouse repaid their confidence. Elected to the Pennsylvania assembly, he served in a variety of important state offices and helped to write the Pennsylvania Constitution of 1776. As engineer of the Pennsylvania Committee of Public Safety, he quickly developed an expertise in military engineering quite removed from his earlier fields of study. He helped to survey strategic points for fortification about Philadelphia and along the Delaware River and to supervise and propose modifications in the manufacture of cannon. Like his fellow scientist-revolutionaries John Winthrop, David Ramsay, and Benjamin Rush, Rittenhouse grappled with the problem of America's dearth of gunpowder and tried to devise a more efficient method of manufacture. He was joined in this effort by Charles Willson Peale, one of the leading painters of his generation. Together the two men went on to attempt two rifle improvements: a telescopic sight—stemming perhaps both from Rittenhouse's astronomical background and Peale's "painter's quadrant"—and a breach box in the stock for carrying bullets and wipers. Many of these experiments met, admittedly, with mixed success. The breach box opened accidentally, spilling its contents; Rittenhouse's idea of rifled cannon proved more ambitious than the pressures of time allowed; the yield of gunpowder fell short of initial expectations. Nevertheless, the character of these efforts testified to Americans' sense that they could comprehend and master at least to some degree military technology, and that such exertions might prove invaluable to the success of the Revolution.[17]

One finds a similar faith and enthusiasm, a sense of alliance between technology and the attainment of liberty, in the efforts of other citizen-engineers. As the brilliant political agitator and social advocate Thomas Paine later observed of the period, "The natural mightiness of America expands the mind, and it partakes of the greatness it contemplates. Even the war, with all its evils, had some advantages. It energized invention and lessened the catalogue of impossibilities." Paine, in effect, suggested the emergence of a new ecology of liberty to complement the nation's agrarian base: the twin forces of America's native grandeur and the necessities of the Revolution were conjoining to produce a progressive national technology. Paine himself exemplified this spirit. A lifelong student of engineering, during the Revolution he suggested to Franklin and

Rittenhouse the use of incendiary iron arrows shot by steel cross-bows to disable the British fleet. In like manner, Sion Seabury proposed to Yale College President Ezra Stiles a rolling breastwork of huge logs by which American soldiers might stalk the British cannon then laying siege to Boston. A Yale student, David Bushnell, was at the same time designing and building a wooden submarine, an awkward-looking device he christened the "American Turtle." According to Bushnell's plan, the submarine would approach an enemy warship underwater and attach a time bomb to its hull, allowing the "Turtle" half an hour to escape. On its maiden voyage in 1776 the submarine successfully reached its quarry—only the detachable screw which was to fix the charge failed to penetrate the copper bottom of the warship. Two later attempts also failed, and Bushnell abandoned the submarine to experiment with floating mines. A generation after Bushnell, Robert Fulton wrestled with the problem of the submarine and devised a torpedo system as a defense against naval attack. Knowledge of the existence of such weapons, he argued, would make further war unthinkable and free men's energies and resources for constructive purposes. In presenting his ideas to President Madison and Congress in 1810, Fulton concluded in an affirmation of faith with which the Revolutionary generation of inventors would have heartily concurred: "Every order of things, which has a tendency to remove oppression and meliorate the condition of man, by directing his ambition to useful industry is, in effect, republican." Thus the challenge of the Revolution bolstered Americans in their technological capacity and strengthened their sense of the reciprocal progress of technology and liberty.[18]

With the achievement of independence, the issue of the place of domestic manufactures and the promotion of technology reached still another critical juncture. Though English tyranny had at last been repulsed, questions of political power, economic policy, and moral virtue—the issues which had swept Americans into Revolution in the first place—continued to consume the nation. The possibility that the republic, at last freed from external oppression, might fail from internal excesses, haunted their minds. There exists an emotional as well as an intellectual continuity between the War of Independence and the period of Confederation: a sense of expectancy, of energy, but also a deep, even anxious concern to hold true to course, a determination not to relax the effort. The powerful lens of republican ideology by which Americans had earlier examined England, they now trained on themselves, and many were dismayed at what they saw. The simplicity, self-sacrifice, and virtue which they had so esteemed in 1775 appeared in danger of degenerating into what one critic called "political pathology."

Particularly alarming in this context was the character of America's foreign commerce. Once hostilities had ceased, British goods poured afresh into the United States, their way smoothed by generous extensions of credit. After years of homespun and austerity, Americans bought the wares of their erstwhile oppressors quickly and eagerly. They found, however, an important outlet for their own materials in the British West Indies closed to their ships. Unable to repay their debts, they soon plunged deep into economic depression, and attempts to climb out of debt through the issuing of paper money only increased the confusion. Critics traced the situation not simply to the breakdown of America's accustomed cycle of trade, but to a new love of luxury, threatening the republican venture itself. "LUXURY, LUXURY, the great source of dissolution and distress, has here taken up her dismal abode," bewailed one observer; "infectious as she is, she is alike caressed by rich and poor." Having lost the war, Britain now appeared to have assumed a more cunning guise in order to seduce by extravagancies the virtue she could not conquer by arms. The Virginian St. George Tucker imagined the British ministry gloating over the situation: "If by any stratagem we can continue to monopolize [America's] trade as heretofore, Britain can suffer no injury whatsoever from the American revolution. —If the profits of her trade center not in her own states, America will ever be indigent and contemptible, while the nation which engrosses her trade will encrease in wealth and pòwer proportionate to her poverty." Once again, then, Americans faced a crisis of economic independence which threatened the entire republican venture. Encouragement of domestic manufactures as a vital defense of republicanism had earlier been a rallying cry in the colonial boycott movements and the Revolutionary War itself. The question now arose whether these efforts should be

fostered as the springboard for a permanent system of manufactures.[19]

On this subject certainly the most familiar contribution is
Thomas Jefferson's celebrated argument that America remain a
nation of farmers, contained in *Notes on the State of Virginia*
(1785). The notion of European political economists that every
nation should manufacture for herself, Jefferson contended, was
mistakenly applied to America, where the abundance of land might
support the industry of all the people. The root of Jefferson's objections were not economic, however, but moral:

> Those who labour in the earth are the chosen people of God,
> if ever he had a chosen people, whose breasts he has made
> his peculiar deposit for substantial and genuine virtue. It is
> the focus in which he keeps alive that sacred fire, which
> otherwise might escape from the face of the earth. Corruption
> of morals in the mass of cultivators is a phænomenon of which
> no age nor nation has furnished an example. It is the mark set
> on those, who not looking up to heaven, to their own soil and
> industry, as does the husbandman, for their subsistance, de-
> pend for it on the casualties and caprice of customers. Depen-
> dance begets subservience and venality, suffocates the germ of
> virtue and prepares fit tools for the designs of ambition.

What, then, did it matter if foreign manufactures might sometimes
be dear? If that was the price of liberty, who could dissent from
Jefferson's injunction, "for the general operations of manufacture,
let our work-shops remain in Europe"?

> It is better to carry provisions and materials to workmen
> there, than bring them to the provisions and materials, and
> with them their manners and principles. The loss by the trans-
> portation of commodities across the Atlantic will be made up
> in happiness and permanence of government. The mobs of
> great cities add just so much to the support of pure govern-
> ment, as sores do to the strength of the human body. It is the
> manners and spirit of a people which preserve a republic in
> vigor. A degeneracy in these is a canker which soon eats to
> the heart of its laws and constitution.[20]

Jefferson's moral stance, his insistence upon the interrelationship of freedom, industriousness, and virtue, was unimpeachable. No one would argue that America should sell her republican birthright for a mess of pottage; nor would anyone deny that agriculture would properly remain the basis of America's economy and the overwhelming occupation of her people for the indefinite future. "The genius of America is agriculture," Richard Wells remarked in 1774, "and for ages to come will continue so." Six years later John Adams observed contentedly, "The principal interest of America for many centuries to come will be landed, and her chief occupation agriculture. . . . America will be the country to produce raw materials for manufactures; but Europe will be the country of manufactures." And in 1787 Benjamin Franklin declared in the same spirit, "The great business of the continent is agriculture. For one artizan, or merchant, I suppose we have at least a hundred farmers."[21]

But if all willingly acknowledged the pre-eminence of agriculture, an increasing number of Americans in the 1780s demanded that the nation stop the rapid sapping of her economic, political, and moral strength through European trade and restore her vigor by promoting domestic manufactures. Against the background of economic and social crisis, these men launched upon a reassessment of America's potentialities and a renewed consideration of the possibilities of republicanism. Building on the arguments and experience of the 1760s and '70s, advocates of manufactures contended with increasing boldness that public virtue and the public good might best be achieved and defended through the establishment of an independent and balanced economy. Their arguments helped pave the way for the establishment of a stronger national government empowered to control trade under the Constitution and to solidify the developing alliance between technology and republicanism.

Appeals for a system of domestic manufactures and regulation of foreign trade took as their starting point the precipitous declension of American fortune and morality in the brief passage since the Revolution. Hugh Williamson, a North Carolina mathematician, scientist, and delegate to the Constitutional Convention, lamented the erosion of industry, economy, temperance, and self-denial which had sustained the nation in her battle for independence. Where the spirit of public virtue had once prevailed, he solemnly declared, now stood "a nation that is more luxurious, more indolent, and more extravagant, than any other people on the face of the earth." Observations on American commerce thus became jeremiads. As another writer preached, "Nations that are remarkable for idleness and sloth, are for the most part prone to luxury, effeminacy, and extravagance. What hasty strides have we not taken since the peace to gain the summit of those refinements! O may the good genius of America now step forth, and inspire her infatuated sons, to make a solemn pause, to consider, and to amend!" An orator at the Petersburgh, Virginia, Fourth of July celebrations in 1787 chose the occasion not to celebrate the nation's progress since independence, but to catalogue her vices and weaknesses and warn of inevitable ruin unless the American people quickly righted their course. The root of present evils, he contended, lay in foreign trade, which "is in its very nature subversive of the spirit of pure liberty and independence, as it destroys that simplicity of manners, native manliness of soul, and equality of station, which is the spring and peculiar excellence of a free government." Only a return to domestic duties, to the joint cultivation of agriculture and manufactures, would restore the republican virtue upon which America's future depended.[22]

In the effort to promote domestic manufactures, the example of colonial resistance and the Revolution became a rallying cry, and various states enacted laws similar to those which bolstered the colonial boycotts and the war effort. As the depression deepened in 1785, Rhode Island, Massachusetts, Pennsylvania, New Hampshire, and New York passed protective tariffs. The expressed purpose of the Massachusetts act was "to encourage agriculture, the improvement of raw materials and manufactures," and also "to discourage luxury and extravagance of every kind." During the next three years a number of states offered special tax exemptions to attract certain manufactures. Pennsylvania went to the extent of offering loans and buying shares to support several enterprises involving steel and cotton, while North Carolina and Massachusetts sought to encourage iron and cotton production, respectively, through land grants.[23]

Similarly, Hugh Williamson called for a renewal of the colonial boycotts and manufacturing associations which preceded and supported the Revolution, in order to restore America's economy and spirit. Instead of "fluttering about in foreign dress," he urged his countrymen to don homespun once again and thus wear the mantle of patriotism as well. And indeed, the same issue of Mathew Carey's journal supporting American manufactures, *American Museum*, which published Williamson's plea, also recorded resolutions from groups in Hartford, Connecticut; Halifax, North Carolina; Richmond, Virginia; and Germantown, Pennsylvania, condemning extravagant purchases of foreign items and bidding Americans to support domestic manufactures by buying simple native dress and goods. A few issues later, another writer in *American Museum* suggested that all federal officers, both civil and military, take a pledge accompanying their oath of office to wear only clothes of American manufacture in the performance of their duties—with fines of a dollar a day for each infraction. These and similar advocates thus argued that simplicity, frugality, industry, patriotism, and other republican virtues stemmed not from agricultural pursuits alone, but depended as well on the promotion of domestic manufactures. Far from being morally enervating and destructive of independence, they replied to Jefferson, manufacturing was essential to liberty and virtue. The true villain was not the honest workman, but the wily foreign merchant. To leave America's workshops in Europe, therefore, meant to induce a fatal cancer into the manners and spirit of the American people.[24]

Even if this argument in favor of manufacturing were accepted, however, another of Jefferson's objections still remained. At the heart of his fervent appeal to a pastoral ideal in *Notes on Virginia* lay the concept of America as "Nature's nation." America derived her virtue and vitality, ran Jefferson's argument, from independent and direct contact with nature. This being so, what place had urban workshops with all their manufactures here? Buttressing this pastoral appeal was the physiocratic doctrine that the only source of real wealth, the only truly productive employment, was agriculture. This axiom, embraced by Franklin as well as Jefferson, was accompanied by a corollary. As the Philadelphia agrarian George Logan stated it, "Every other kind of work which has for its object the preparation, alteration, or transport of such productions may be more or less necessary, but they are not productive."[25]

Advocates of American manufactures, then, needed to remove the stigma that their cause was in some respect unproductive, derivative, even unnatural before they could fully integrate manufactures into the ideology of republicanism. Such a reformulation of attitude and value, of course, could hardly be achieved overnight. In fact, as we shall see, the question of the relationship of manufactures and large-scale technology generally to nature would continue to concern Americans throughout the nineteenth century —indeed, up to our own time. What is important to note here is that proponents of manufactures were sensible of this onus and took pains to discharge it. In making this attempt, they sketched the outlines of an argument that Whig orators a half century later would master: they insisted that manufactures, as well as agriculture, harnessed natural resources and fulfilled nature's purpose; and that in the face of such potentialities, agrarian critics were less vigilant shepherds than carping aesthetes, melancholy Jaqueses in an industrial Arden.

Thus, an anonymous "plain, but real friend to America" challenged agrarians' objections to manufactures in the *American Museum*. In insisting upon farming as the sole natural occupation, he suggested, they construed the possibilities of nature in the New World too narrowly. Precisely because nature had been so "profusely liberal" to America, it became her to use her gifts to the utmost. This meant engaging not solely in agriculture but in manufactures as well. Essentially, the author justified promotion of domestic manufactures by arguing for a divine utilitarianism:

> Nature does nothing in vain. Her operations are regulated by the nicest and best rules. What she gives us in our own country, we may rest assured, if rightly used, will be found to be the best for us. Conduct not yourselves, therefore, my countrymen, as if you believed that nature bestowed on one country what ought to be given to another, which absurd idea

would be chargeable on you, for your spurning at her gifts, by either wholly neglecting them, or sending them abroad to be manufactured. How contrary this to the dictates of common reason! Be wise for the future. Learn to prize the numerous blessings which God and nature have favoured you with.

For Americans to import finished goods instead of making their own was thus a violation of the economy of nature, a rejection of her beneficence, and a flouting of America's destiny. True frugality and industry meant the pursuit of both agriculture *and* manufactures. Nature, technology, and republicanism wonderfully cohered. Only sentimental pastoralism, the writer suggested, blinded critics from perceiving that nature's purposes were not fulfilled until her powers were fully utilized. "It is not hills, mountains, woods, and rivers, that constitute the true riches of a country. It is the number of industrious mechanic and manufacturing as well as agriculturing inhabitants." America, in fact, was not immune to the laws of natural and political economy which governed the rest of the world, and therefore to gaze dreamily at the scenery while the fortune and morals of the nation declined was to court disaster.[26]

Even as Americans discussed the advantages of promoting manufactures, the character of manufacturing abroad was being radically transformed. Great Britain's cotton industry, in particular, was by now well launched upon a revolutionary series of innovations in mechanization and production, and other industries were beginning to experience similar effects. Dramatic changes came in a pattern of challenge and response, with innovations in one stage of manufacture demanding corresponding improvements in the other stages. Thus as John Kay's fly-shuttle (1733) achieved widespread use in the 1750s and '60s, it triggered a series of developments in spinning: first, carding machines in the 1750s; then James Hargreave's jenny (c. 1765; patented 1770); Richard Arkwright's water frame (1769); and Samuel Crompton's mule (1779), which combined the spinning action of the jenny with the drawing action of the water frame. The rate of innovation so increased that the spinning wheel—which had taken centuries to replace the distaff—was superseded in a decade, and the jenny itself became obsolete within a generation. Improvements in weaving were stimulated literally to take up the slack, and by 1787 Edmund Cartwright had invented a crude version of the power loom. The pattern of challenge and response was then extended, in turn, to the preliminary and finishing processes of cotton manufacture. The effect of these innovations on production was immediately apparent. British imports of raw cotton, 2½ million pounds in 1760, leaped to 22 million pounds by 1787. In another half century, it would reach 366 million pounds.[27]

This revolution in mechanization, symbolized by cotton manufacture, was accompanied by a revolution in power and energy. In less than a century steam engines fired by coal had advanced from Thomas Savery's steam pump of 1698 to Thomas Newcomen's piston-equipped atmospheric steam engine of 1705 to the series of improvements by James Watt, beginning with the installation of a separate condenser in the 1760s. Windmills and water wheels would continue to supply the majority of British industries' power into the early nineteenth century, and in America, where mill seats were more plentiful and coal deposits more inaccessible, they would remain dominant even longer. Yet almost immediately steam became symbolic of the emergent technological order, to be celebrated as such throughout the nineteenth century, and steam-powered spinning mules were in use in England by the early 1790s. The boast of James Watt's partner, engine manufacturer Matthew Boulton, to James Boswell in 1776 would thrill generation after generation: "I sell here, Sir, what all the world desires to have— POWER."[28]

What were the implications of this new power for republicanism? Did Americans regard these new forces of energy and production suspiciously as they did political power, as threats to their liberty? Quite the opposite. Such marvels appeared to them victories of the human mind and spirit, promising a grand new era of progress in which America would stand in the forefront. Even Thomas Jefferson, despite his desire to see America's workshops "remain in Europe," grew elated at the spectacle. Visiting John and Abigail Adams in England in 1786, a year after the publication of *Notes on Virginia*, he accompanied them to a new mill at Black-friars Bridge on the Thames, powered by huge Boulton and Watt steam engines. While Jefferson was not allowed to inspect the

Jefferson's Mill at Shadwell: A nineteenth century view.

machinery closely owing to English precautions against the exportation of industrial secrets, he came away greatly impressed by the
visit. To an American friend, he wrote excitedly:

> I could write you volumes on the improvements which I find
> made and making here in the arts. One deserves particular
> notice, because it is simple, great, and likely to have extensive
> consequences. It is the application of steam as an agent for
> working grist mills. I have visited the one lately made here.
> It was at that time turning eight pair of stones. It consumes
> 100. bushels of coal a day. It is proposed to put up 30. pair of
> stones. I do not know whether the quantity of fuel is to be
> increased. I hear you are applying this same agent in America
> to navigate boats, and I have little doubt but that it will be
> applied generally to machines, so as to supersede the use of
> water ponds, and of course to lay open all the streams for
> navigation. We know that steam is one of the most powerful
> engines we can employ; and in America fuel is abundant.[29]

Jefferson's concluding remarks, in which he speculated upon applications of the new technology to the American scene, were characteristic of his generation. Americans were particularly eager to probe the special meaning technology might have for their lives and the development of their society.

By turning briefly to Jefferson's own experience we may gain additional insight as to how Americans conceived the relationship between unfolding technological developments and republicanism. Jefferson, it should be noted, was never opposed to simple, household manufactures, only to large-scale enterprises yielding special benefits to a few and spawning a debased proletariat. And, of course, under the pressure of the American embargo begun in 1807 and the War of 1812, he reluctantly concluded that America had no choice but to manufacture for herself.[30] However, long before Jefferson was forced to that position, he not only delighted in the use of various mechanical contrivances and labor-saving machines around his house and farm at Monticello and devised improved threshing machines and plows, but—what is often less emphasized —was an enthusiastic plantation manufacturer as well. Upon his brief "retirement" from public affairs in 1794, he returned to Monticello to become not only a farmer but a nail-maker. Beginning with a single fire and making nails by hand, he bought a cutting machine in 1796, added two more fires, and achieved an output of 10,000 nails a day. A dozen slave boys from ten to sixteen years old worked at the nailery under the blacksmith's direction, and Jefferson himself supervised the project and worked frequently at the forge. The scene alters our conventional image of Jefferson. As his biographer Merrill Peterson remarks, "Here was no pastoral Eden but belching smoke and clanging hammers."[31]

Jefferson went on to build a toll or grist mill and a manufacturing mill at his Shadwell plantation on the Rivanna River and a textile manufactory for his household at Monticello. In each case he pursued the technical details of construction thoroughly and eagerly, taking pains to obtain the most efficient machinery available. Describing his textile establishment to Thaddeus Kosciusko upon the outbreak of the War of 1812, he wrote:

> Our manufacturers are now very nearly on a footing with those of England. She has not a single improvement which we do not possess, and many of them better adapted by ourselves

to our ordinary use. We have reduced the large and expensive machinery for most things to the compass of a private family, and every family of any size is now getting machines on a small scale for their household purposes. Quoting myself as an example, and I am much behind many others in this business, my household manufactures are just getting into operation on the scale of a carding machine costing $60 only, which may be worked by a girl of twelve years old, a spinning machine, which may be made for $10, carrying 6 spindles for wool, to be worked by a girl also, another which can be made for $25, carrying 12 spindles for cotton, and a loom, with a flying shuttle, weaving its twenty yards a day. I need 2,000 yards of linen, cotton and woolen yearly, to clothe my family, which this machinery, costing $150 only, and worked by two women and two girls, will more than furnish.

Two years later Jefferson and his son-in-law Thomas Randolph together had expanded their facilities to four jennies with a total of 112 spindles and carders and looms to accompany them.[32]

For Jefferson, then, the significance of increased mechanization and new forms of power was not that they would lead to "dark satanic mills" in America, but on the contrary, that they would allow America to manufacture for herself while circumventing these evils. The very machines and sources of energy which we now see to have compelled a concentration of production in large factories, Jefferson construed as permitting increased *decentralization* of production, as offering household manufactures a new viability and productivity. Remaining within an essentially agricultural context instead of clustered in cities, such manufactures would not threaten, but rather enhance, the independence and virtue of American society. Within such a vision, reliance upon the labor of "a few women, children and invalids, who could do little on the farm" was not a blot but a humane and useful employment. The new technological developments of the late eighteenth century which would swell into what we call the Industrial Revolution, in short, did not prompt Jefferson to redouble his opposition to the promotion of American manufactures; rather they encouraged him to moderate his position, to see in the new technology a welcome ally in the republican enterprise.[33]

Before Jefferson ever embarked upon his ventures in manu-facturing, however, active proponents of American manufactures, less suspicious than he of their possible baneful effects, had seized upon the implications of contemporary technological innovations for American republicanism. Manufacturers gathered news of British achievements eagerly, and restrictions against exportation of British machines and specifications or the emigration of skilled workmen only stimulated their appetites and spurred their efforts. Contravention of British law became unofficial American policy; and the smuggling of machinery and attraction of mechanics turned into a "fine art." We have already seen how the United Company of Philadelphia imported a spinning jenny as early as 1775; in the late 1780s Americans redoubled attempts to introduce British tex-tile machines and other innovations into this country and so estab-lish domestic manufactures on a firm footing. In 1786 Hugh Orr, backed by the Massachusetts legislature, engaged two Scotsmen, Robert and Alexander Barr, to construct and demonstrate what was reportedly the first jenny and stock card made in America. At the same time Thomas Somers, an Englishman living in America, visited English cotton manufactories and returned with descrip-tions and models to build a primitive form of Arkwright's water frame. Inspired by Somers' work, a group of Massachusetts mer-chants headed by George Cabot established a cotton factory at Beverly in 1787. When the new President, George Washington, visited their plant two years later, he admiringly described in his diary "new invented carding and spinning machines" and looms with spring shuttles, concluding, "In short, the whole seemed per-fect, and the cotton stuffs which they turn out excellent of their kind." These and similar attempts were climaxed in 1790 by the most celebrated act of industrial espionage in American history, when Moses Brown brought Samuel Slater to Pawtucket, Rhode Island, to reduplicate from memory Richard Arkwright's water frame and other machines whose construction he had overseen in England, thus establishing the most advanced spinning machinery in America.[34]

All these entrepreneurs labored under immense obstacles: the difficulty of obtaining reliable machines or models, the lack of trained engineers and workers, as well as a shortage of capital, and the continual pressure of British competition. Yet they persisted through numerous trials and failures, confident that success would eventually crown their efforts. The correspondence amassed by Alexander Hamilton in preparation for his famous *Report on Manufactures* of 1791 reveals an extensive and rapidly expanding American industry which gave substance to the rhetoric of economic independence. In response to Hamilton's inquiry, Hartford Woolen Manufactory's Elisha Colt wrote, "The present use of Machines in England give[s] their Manufacturers immense advantage over us—This we expect soon to remedy." His terse statement reflected the determination of many manufacturers and their expectation of the transforming effects of the new technology.[35]

Thus, in the late 1780s the example of British technological innovations notably stimulated Americans' appetite for domestic manufactures. Recent advances, particularly in mechanization, extended the promise that the United States might at last resolve the dilemma she faced in establishing a republican economy and society: she could remain a nation of farmers with all the sanctity and virtue of that vocation and at the same time, through domestic manufactures, protect her republican condition from external corruption. While able-bodied men tilled the soil, manufactures would be produced by machines, assisted by marginal laborers: women, children, and the aged.

The modern reader shudders at the specter of children, women, elders, and invalids toiling in the factory. But the prospect did not seem harsh to a society in which they were expected to work at some task in any case. We have already seen the equanimity, even enthusiasm, with which Jefferson regarded such employment. In a similar vein, George Washington wrote in 1789, "Though I would not force the introduction of manufactures, by extravagant encouragements, and to the prejudice of agriculture, yet I conceive much might be done in that way by women, children, and others, without taking one really necessary hand from tilling the earth." "The husbandman himself," wrote Hamilton two years later in his *Report on Manufactures*, "experiences a new source of profit and support, from the increased industry of his wife and daughters, invited and stimulated by the demands of the neighboring manufactories." And in 1794, urging Congress to pass a protective tariff to encourage American manufactures, David Humphreys was moved to sing:

> To useful arts a nation's aid direct,
> Create new fabrics and the old protect;
> Bid fire and vapour hear the artist's call,
> And water labour in its forceful fall,
> Ingenious engines wondrous works perform,
> The hungry nourish and the naked warm!
> Teach little hands to ply mechanic toil,
> Causing failing age o'er easy tasks to smile;
> With gladness kindle rescu'd beauty's eye,
> And cheek with health's inimitable dye;
> So shall the young, the feeble find employ,
> And hearts, late nigh to perish, leap for joy!

In the hands of writers like Hamilton and Humphreys, the language of republican industry and virtue melded sweetly into the sound of the factory bell. Both men attempted to realize their visions by founding model factory towns relying heavily on the labor of women and children.[36]

While other Americans grew excited at the prospect of revolutionary new applications of marginal labor, Delaware's Oliver Evans went even further: More than perhaps any other inventor in the eighteenth century he succeeded in substituting machines for human labor entirely. On the banks of Redclay Creek in the 1780s he mechanized the complicated craft of the miller, building a mill in which grain moved from the time of its unloading through its transformation into processed flour by a series of conveyors in continuous line production without any human assistance. Well over a century before Henry Ford and his associates applied the continuous automatic conveyor to automobile production in 1914, Evans had developed the principle of the assembly line. His achievement stood as the period's most dramatic example of the way in which new technological developments promised to transform American manufactures while avoiding the creation of a degraded proletariat.[37]

The clearest and most sustained statement at this time of modern technology's special significance for the republican enterprise came on August 9, 1787, when the Philadelphia merchant Tench Coxe delivered the inaugural address before the Pennsylvania Society for the Encouragement of Manufactures and the Useful Arts. Though only thirty-two years old, Coxe had been extensively and actively concerned with the development of domestic manufactures. His father had participated in the nonimportation movement following the Stamp Act in 1765, and Coxe himself as a young man of twenty had subscribed to the United Company of Philadelphia for Promoting American Manufactures. A member of the Annapolis Convention of 1786, which prepared the way for the Philadelphia Constitutional Convention the following year, he was one of a growing number of Americans who argued that the Articles of Confederation should be redrawn. In May 1787, three days before the Constitutional Convention began, Coxe gave a paper before the Society of Political Enquiries at the home of Benjamin Franklin in which he forcefully argued for the establishment of a nationally coordinated economy of agriculture, manufactures, and commerce, which would direct the United States toward a balanced self-sufficiency. Three months later in his speech to the Pennsylvania Society, Coxe made clear the vital part to be played by modern technology in realizing his ideal.[38]

In his opening remarks to the Society, Coxe gestured grandly to "the AUGUST BODY," the Constitutional Convention, then laboring through the summer heat; yet for Coxe the assurance of American republicanism depended not alone upon the revision of political institutions but also upon the cultivation of domestic manufactures. Such sentiments had become widespread by 1787, even though resistance to manufactures still remained. The importance of Coxe's speech, however, was that he revealed the special implications of contemporary technology for America; he showed how the machine might be harnessed to a republican civilization. New machines and sources of power, Coxe contended, made objections to American manufacturing on grounds of the scarcity and dearness of labor and raw materials and the work's harmful effects upon laborers no longer tenable. The vision he heralded was the American ideal of a workshop without workers: an automated factory in which employees would perform light preparatory and supervisory tasks while the bulk of the work went forward by machines. As Coxe described the prospect to his audience: "Factories, which can be carried on by water-mills, wind-mills, fire, horses, and machines ingeniously contrived, are not burdened with any heavy expence of boarding, lodging, clothing, and paying workmen; and they multiply the force of hands to a great extent, without taking our people from agriculture." Recent labor-saving inventions, such as those developed by the English textile industry, Coxe foresaw, promised to benefit America far more than Europe. The nation's scarcity of labor, no longer a liability to manufactures, on the contrary became an asset, an incentive toward mechanization, whereas in England, where labor was more plentiful and hence cheaper, the advantages were not as great. Coxe cited the example of a European factory which, with "a few hundreds of women and children ... performs the work of twelve thousand carders, spinners, and winders," and enthused over the possibilities of power-driven machinery for America: "Perhaps I may be too sanguine, but they appear to me fraught with immense advantages to us, and full of danger to the manufacturing nations of Europe: for should they continue to use and improve them, as they have hitherto done, their people must be driven to us for want of employment: and if, on the other hand, they should return to manual labour, we shall underwork them by those invaluable engines." Savoring the irony, Coxe was telling his audience that England had, by her own artful devices, surrendered her competitive advantage in manufactures to America.[39]

The effect of the new technology, as Coxe described it, was to reverse not only economic objections to manufactures, but moral ones as well. The overwhelming character of American life would continue to be agricultural; as for manufactures, "Horses, and the potent elements of fire and water, aided by the faculties of the human mind (except in a few healthful instances), are to be our daily labourers." Machine-powered factories would serve in effect as republican institutions and provide a strong moral antidote to elements of dissipation and corruption. By employing those poor unsuited for other work, factories would relieve discontent and potential disorder. As Coxe observed, "A man oppressed by extreme want, is prepared for all evil: and the idler is ever prone to wickedness: while the habits of industry, filling the mind with honest thoughts, and requiring the time for better purposes, do not leave leisure for meditating or executing mischief." Similarly, native manufactures would combat luxury. Coxe pointed to America's "untimely passion for European luxuries as a malignant and alarming symptom, threatening convulsions and dissolution to the political body." This debauched taste could also be cured by manufacturing simple domestic goods. American manufactures thus emerged at the conclusion of Coxe's address as the beacon of republicanism: "It will lead us, once more, into the paths of virtue, by restoring frugality and industry, those potent antidotes to the vices of mankind, and will give us real independence by rescuing us from the tyranny of foreign fashions, and the destructive torrent of luxury."[40]

In Coxe's address the union of technology and republicanism was complete. He defended American manufactures not only as a bulwark against European luxury but also as a positive agent of domestic social virtue. The earnestness with which he warned against threats to the American social and political order suggests that the republican stress upon simplicity, frugality, industry, and virtue had become, as Gordon Wood has observed, "an ideology of social stratification and control." Many advocates of governmental revision in 1787 were concerned not only with the need for a stronger national government empowered to regulate trade and thus establish a national economic policy, but also with power and influence to check disturbing symptoms of social discord. Complaints of a corruption of virtue, an increase of luxury by which people aped their betters, a corrosion of manners and social deference, and a general breakdown of authority circulated throughout the 1780s, reaching a climax in the winter of 1786–87 when two thousand indebted farmers in western Massachusetts under the leadership of Daniel Shays rose in rebellion against the state government. Shays's Rebellion sent shock waves throughout the country and brought the sense of social and governmental crisis to a head. If republicanism were allowed to degenerate into social leveling and anarchy, social conservatives feared, then America's worst critics would stand vindicated. Dependence upon the intrinsic virtue and restraint of the people would no longer suffice. Social and political instrumentalities had to be devised to ensure the leadership of America's "natural aristocracy" and, in the words of John Dickinson, to protect "the worthy against the licentious."[41]

In his speech to the Pennsylvania Society, Coxe suggested how mechanized factories might serve in such a task; and, as we shall see later in the history of Lowell, Massachusetts, his image of the factory as a republican institution and symbol of social order proved prophetic. At the same time other Americans concerned with strengthening the bonds of society against dissension prescribed related remedies. Thus Benjamin Rush, who had publicly urged the promotion of domestic manufactures as early as 1775 and who as president of the Pennsylvania Society suggested Tench Coxe give his address, published in 1786 his own proposal for "the Mode of Education Proper in a Republic." With the achievement of independence and the establishment of a republic, according to Rush, came new duties which American education had a responsibility to inculcate. "Our schools of learning, by producing one general, and uniform system of education, will render the mass of people more homogeneous, and thereby fit them more easily for uniform and peaceable government." Republican instruction, Rush maintained, should teach Americans to love their country above all else—family, friends, and property—and to regard themselves as her protectors and servants. "I consider it possible to convert men into republican machines. This must be done, if we expect them to perform their parts properly, in the great machine of the government of the state." Rush's metaphorical use of the word "machine" derived its power not only from Newtonian mechanics but also from the new technological innovations with which Rush was intimately associated; it revealed the emerging alliance of technology and other republican institutions around a common purpose of social order and control. The characteristic qualities of mechanization—regularity, uniformity, subordination, harmony, efficiency—appeared to offer a model for government and society in general. If Tench Coxe's factory was to be a school of republicanism, teaching industry and virtue, Rush's ideal school would be in its own way a factory, producing model "republican machines."[42]

The Constitutional Convention in 1787 similarly represented an attempt to institutionalize republicanism. Delegates sought to protect and consolidate the victories of the Revolution in a stronger national government while curbing its democratic excesses. In the language of *The Federalist* they proposed to establish a political system in which public "passions" would be "controlled and regulated" and disorder restrained by "an efficient national government" of "the best men." Almost all would have agreed with George Washington's rueful observation to John Jay the previous year: "We have probably had too good an opinion of human nature in forming our confederation. Experience has taught us, that men will not adopt and carry into execution measures the best calculated for their own good, without the intervention of a coercive power." As in Rush's proposal for American education, the overriding metaphor for the structure of government within the Constitution itself was a machine, which through a system of "checks and balances" harmoniously regulated itself and channeled the energies of the people. The image undoubtedly reflected the delegates' and the age's conception of an orderly universe, but it suggested as well the common purpose of technology and republican government.[43]

In addition to the framers' common concern in establishing republican controls, the Constitution marked the culmination of the American effort to establish an independent national economy, which had begun with the nonimportation agreements and campaign for domestic manufactures in 1764-65. Again and again, in the face of each new crisis, preceding the Revolution, during the War itself, and in the period of Confederation, Americans responded by calling for the promotion of domestic manufactures and increased technological development; repeatedly they insisted that they would not be truly free, that republicanism would not be secure, until they were freed from corrupting contact and dependency upon European manufactures and trade. However, when in the 1780s individual state governments attempted to deal with the challenge of foreign commerce, they produced only a crazy-quilt of commercial regulations and an alarmingly unstable economy. The necessity for the creation of a national government with power to regulate commerce and hence establish national economic policy grew increasingly clear; and trade conventions at Mount Vernon in 1785 and Annapolis in 1786 led like stepping stones to the Constitutional Convention. Without Congressional authority over interstate commerce, tariff restrictions by individual states might have continued to block the development of a national market. The result would have been to have discouraged regional specialization and large-scale production which, supporters of the new Constitution foresaw, new technology was beginning to make possible. The Constitution thus provided an essential foundation for American economic and technological growth.[44]

Constitutional advocates were further induced to think "continentally," as Alexander Hamilton put it, and their sense of the possibilities of the republic was enlarged by recent improvements in transportation and communications among the states. Following the Revolutionary War, numerous states, private companies, and individuals launched a host of projects to facilitate transportation: clearing streams and rivers; building roads, bridges, and canals; carrying mail in stagecoaches from New Hampshire to Georgia; even succumbing to the French mania for ballooning. At the same time John Fitch and James Rumsey vied with one another to develop a practicable steamboat; and Fitch's successful demonstration on the Delaware River in August 1787 was witnessed by a number of delegates to the Constitutional Convention.[45]

Many of the opponents of the Constitution lacked such a continental vision. They drew on Montesquieu and the most respected political theory of the age to argue that republican government was possible only within a relatively limited area with a small and homogeneous population. The American people were too diverse and a central government would be too distant, they contended, for it to work effectively; the necessary outcome of the federal system would be either collapse or, worse, despotism. As one Antifederalist wrote, "The idea of an uncompounded republick, on an average one thousand miles in length, and eight hundred in breadth, and containing six millions of white inhabitants all reduced to the same standard of morals, of habits, and of laws, is in itself an absurdity, and contrary to the whole experience of mankind."[46]

It did not seem absurd to the Federalists, however, and they endeavored to reverse the thrust of theorists from Plato to Montesquieu that republican government could not be extended over a vast area. No republic, they asserted, no matter how small, really contained a homogeneous people, but a large federal republic could better achieve governmental stability and contain the violence of faction than the individual states. Moreover, improvements in transportation and communication promised to bind the states securely together and promote a sense of unity; as such they would provide a powerful check against regional dissension and fragmentation. The most effective refutation of the Antifederalist argument was presented by James Madison in *The Federalist*, and an important aspect of his argument was that "the intercourse throughout the Union will be facilitated by new improvements":

> Roads will everywhere be shortened, and kept in better order ... an interior navigation on our eastern side will be opened throughout, or nearly throughout, the whole extent of the thirteen States. The communication between the Western and Atlantic districts, and between different parts of each, will be rendered more and more easy by those numerous canals with which the beneficence of nature has intersected our country, and which art finds it so little difficult to connect and complete.[47]

That national unity which nature had intended, Madison suggested, the technological arts would fulfill. It was a vision that would be extended and amplified throughout the nineteenth and into the twentieth century as new technological developments suggested new possibilities: from the National Road, canals, railroad, and telegraph, to the automobile, airplane, radio, and television. Technology would help to preserve the Union and make the republican experiment possible.

Thus the seeds of the great era of American industrial and technological growth in the second half of the nineteenth century were planted much earlier and cultivated meticulously through the Revolutionary period to the ratification of the Constitution. That document which many historians have taken to mark the beginning of American industrial consciousness and have regarded as uniquely prophetic, Hamilton's *Report on Manufactures* of 1791, was in fact the culmination, gathering together and summarizing arguments that had been developing for a generation. An apologia for manufactures and a justification for expanding America's protectionist policy, the *Report* was distinguished more in the boldness of its specific recommendations for protective tariffs and bounties than in the originality of its thesis of the importance of a balanced economy. The popular image of Hamilton, encouraged by some scholars, as a Machiavellian figure who smuggled in the blueprints for American industrialization behind the backs of an idyllic and unwary nation of farmers wildly exaggerates not only Hamilton's personal character but the thought and character of the rest of the nation as well. The promotion of American manufactures and modern technology generally was not a conspiracy hatched in private but a campaign relentlessly urged in public, under the banner of American republicanism, for twenty-five years before Hamilton's *Report.* By 1791 their position in the nation's ideology and culture was established.

Modern technology thus appeared to have rescued the young nation from the horns of one republican dilemma; but within a generation, by the second quarter of the nineteenth century, it posed another. Enlisted in the defense of republican virtue, modern machinery threatened to undermine older values by its very success. By tapping natural resources and multiplying the results of American labor, the new technology provided the nation with unprecedented and ever-increasing wealth. America had resisted the lure of British luxuries, but could the country withstand the temptations of her own new prosperity? Writing to Jefferson in 1819, John Adams wrestled with the problem of preserving republican virtue: "Will you tell me how to prevent riches from becoming the effects of temperance and industry? Will you tell me how to prevent riches from producing luxury? Will you tell me how to prevent luxury from producing effeminacy intoxication extravagance Vice and folly?"[48] Adams's concern over the perils of prosperity was reiterated again and again. America was exposed to special danger as a land of abundance, thundered Lyman Beecher in 1829: "The power of voluntary self-denial is not equal to the temptation ... and no instance has yet occurred, in which national voluptuousness has not trod hard upon the footsteps of national opulence, destroying moral principle and patriotism, debasing the mind and enervating the body, and preparing men to become, like the descendants of the Romans, effeminate slaves." Though Beecher firmly believed in the importance of native manufactures, he warned that the nation must cling with redoubled strength of purpose to the moral power of Christianity, or else plunge into "the fire which is destined to consume us."[49]

The dilemma in which Beecher and Adams found themselves was not altogether new. Their forefathers in seventeenth-century New England had similarly striven for prosperity while fearing its results. For, as scholars since Max Weber have repeatedly observed, the Puritan ethic was Janus-headed: it demanded diligence in a calling useful to both the individual and society, yet warned against succumbing to the temptations of success, whether sensuality, sloth, or profiteering. As John Cotton declared in *The Way of Life* in 1641, "If thou beest a man that lives without a calling, though thou hast two thousands to spend, yet if thou hast no calling, tending to publique good, thou art an uncleane beaste." It is this insistence upon the primacy of the "publique good" to which Gordon Wood refers when he declares that both Puritanism and republicanism were essentially anti-capitalistic and attempted to restrain individualistic drives that threatened community life.[50] However, the ideal proved impossible to sustain in fact. Prosperity, a blessing to the virtuous, became a source of temptation to the ungodly. As Perry Miller has shown, beginning in the second half of the seventeenth century and lasting throughout the eighteenth, Puritan ministers released a steady torrent of jeremiads, cataloguing the decline of godly community in apostasy, pride, sensuality, strife, and excessive devotion to worldy concerns. If these jeremiads did not induce lasting repentance and deliverance from the perils of prosperity, they at least offered moments of confession and rituals of obeisance to the standards of the Puritan fathers, even while their sons marched steadily into more worldy spheres.[51]

The ambiguous relationship between prosperity and public virtue with which Puritans wrestled continued to plague Americans into the nineteenth century, as Adams's passionate outcry demonstrates. Yet the extraordinary productive capacities made possible by innovations in machine technology added a powerful new factor which altered the whole context of discussion, challenging basic assumptions concerning prosperity and republican progress. In his letter to Jefferson, Adams expressed an essentially cyclical view of history by which the very products of virtue led societies to luxury, vice, and decay. However, Adams himself was willing at other moments to acknowledge that advances in knowledge and technology marked permanent contributions in the progress of mankind and the achievement of liberty; in another letter to Jefferson, he reiterated the prefatory remarks to his *Defence of the Constitutions* (1787): "The Arts and Sciences, in general, during the three or four last centuries, have had a regular course of *progressive* improvement. The Inventions in Mechanic Arts, the discoveries in natural Philosophy, navigation and commerce, and the Advancement of civilization and humanity, have occasioned Changes in the condition of the World and the human Character, which would have astonished the most refined Nations of Antiquity." Jefferson agreed. "Science is progressive," he wrote to Adams, "and talents and enterprise on the alert." Knowledge, technology, and republicanism, he believed, would in the sweep of history collectively advance.[52]

The new promise of prosperity tempted others to go even further and to suggest what was unthinkable to Adams: luxury was not necessarily an evil, but in the proper context might prove socially serviceable as well as individually rewarding. One of the first to propose this notion was Benjamin Franklin. As early as 1784 he wrote, "Is not the Hope of one day being able to purchase and enjoy Luxuries a great Spur to Labour and Industry? May not Luxury, therefore, produce more than it consumes, if without such a Spur People would be, as they are naturally enough inclined to be, lazy and indolent?" Here was a redefinition of virtue—or at least a recognition of the constructive ends of vice—in which the collective good might be achieved through the pursuit of personal gain. Such a relaxation of Puritan and republican suspicions of luxury paved the way for the coming of industrial capitalism. As the nation advanced into the nineteenth century, the sense of political crisis receded, and opportunities for wealth multiplied, the emphasis upon self-control changed to one of self-fulfillment; the older industrious ethos was transformed into an industrial one. The influential early proponent of American manufactures, Mathew Carey, indicated the changing cultural climate when he argued, "With the generality of mankind, the effect of extreme adversity, is, to harden the heart, and curdle the milk of human kindness in the breast; whereas on the other hand, prosperity generally expands the heart, and leaves it accessible to the suggestions of benevolence and beneficence." Far from a source of temptation, prosperity, according to Carey, was a boon to both body and spirit.[53]

Not everyone shared Carey's genial assurance, of course; and recent historians have pointed to the strain of anxiety and fear of decline which lay beneath the ebullient optimism of the first half of the nineteenth century. In moments of crisis America's moral spokesmen continued to reiterate their warnings of the dangers prosperity posed for republicanism. Like the jeremiads of old, such lamentations offered at once confession and purgation; they allowed nineteenth-century Americans to pay homage to their fathers' more austere standards of virtue, even as the descendants exploited the new possibilities of industrial capitalism.[54] One sees this divided response at work in an issue such as the 1832 debate over the recharter of the Second Bank of the United States. As Marvin Meyers has brilliantly shown, Jacksonians conceived their attack upon the Bank as an attempt to slay "the Monster" which, under the control of foreign interests and "the rich and powerful" at home, threatened to subvert the republican virtue and liberty of the people. For Jackson and his followers, the Bank became the latest incarnation of the spirit of luxury. They feared that its paper money, by encouraging a delusive wave of speculative credit, would inspire an "eager desire to amass wealth without labor" and hence seduce men from "the sober pursuits of honest industry." Yet as Meyers concludes, "the Jacksonians were at once the judges and the judged," torn between their image of the eighteenth-century republic and the prospect of the great economic and social opportunities of the nineteenth century. For these dual impulses the Bank proved an ideal target. Its destruction simultaneously provided a symbolic victory for republican ideology and released institutional restraints over the development of laissez-faire capitalism.[55]

Fears of the baneful effects of wealth and luxury thus certainly persisted, but what is most striking is the degree to which they were placated and subdued by the promise of technological progress. Jackson might castigate the Monster Bank for whetting the appetite for "wealth without labor" but never the new labor-saving devices or the rapid technological progress in manufactures and internal improvements.[56] These were achievements in which the whole country stood to gain. Thus Jackson could welcome their prosperity without any sense of uneasiness. The heady spirit of technological progress prompted others, including many of Jackson's Whig opponents, to go even further and to proclaim that luxury itself under the aegis of technology no longer threatened the republican experiment. Appropriately enough, this new conviction was most clearly and ringingly expressed to celebrate the erection of what amounted to America's first cathedral to the new machine age, the New York "Crystal Palace," at the nation's first world's fair in 1853. For the occasion, the Reverend Henry W. Bellows, one of the most influential Unitarian leaders in the country, delivered an oration on "The Moral Significance of the Crystal Palace." Declared Bellows, "Luxury is debilitating and demoralizing only when it is exclusive. . . . The peculiarity of the luxury of our time, and especially of our country, is its diffusive nature; it is the opportunity and the aim of large masses of our people; and this happily unites it with industry, equality, and justice." By ostensibly providing the means of wealth for all, technology made luxury safe for democracy.[57]

The impact of technology upon American values during this period is reflected in popular typographical depictions of the figure of Liberty. In a number of cases, Liberty stands flanked by symbols of manufacture or agriculture, often with locomotives or steamboats in the background, a maiden of classical virtue but squarely situated in the technological milieu of the nineteenth century. Often she holds a cornucopia as well, and thus Liberty merges with another figure, Prosperity.[58]

This is not to suggest that Americans in any sense inclined to repudiate liberty as a national value or even that they felt a divided allegiance between liberty and prosperity. The point is rather the opposite: To an extent unthinkable a generation earlier, Americans after the War of 1812 defended the merit of their institutions and appealed to the world to judge them according to the standard of prosperity. Speaking in Congress on the Greek revolution in 1824, Daniel Webster made this point quite explicit. America's republican venture, he proclaimed, had yielded "the

greatest possible prosperity" and had earned "distinction and respect among the nations of the earth." Webster commended America's republican example to others in terms of its material accomplishments: "We shall no farther recommend its adoption to other
nations, in whole or in part, than it may recommend itself by its
visible influence on our own growth and prosperity." Republican
virtue, clearly, need not be its own reward. For Webster, the
cornucopia was Liberty's true torch to the world.[59]

From the 1820s onward, with the development of new or improved machines, modes of production, communication, and transportation, and particularly after the introduction of the railroad,
Americans increasingly identified the progress of the nation with
the progress of technology, and native inventors became the objects of a national cult. Patriots boasted that their national pantheon
consisted not of martial heroes, but of peaceful benefactors of mankind, such as Fulton (the American Archimedes), Morse (the
American Leonardo), Franklin, and Whitney.[60] By the early 1850s
fascination with technology so pervaded the culture that, observing
a boys' school during a period of undirected drawing, the Swedish
novelist Fredrika Bremer discovered most of the children sketching
on their slates "smoking steam-engines or steam-boats, all in movement." At this same time lunatics in the New York State Asylum
were busily working on plans for leather frying pans and elliptical
springs to cushion Niagara Falls. The nation as a whole was seized
by a mania for invention. A perfectionist gleam shone in American
eyes before the machine which far surpassed any mere interest in
mechanics. Taken at face value, the countless panegyrics composed
on the sublimity of textile mills or the fabulous grandeur of locomotives appear ludicrous; seen as the utterances of ascetic and practical Yankees, they are nonsensical. They are comprehensible only
when it is understood that in the nineteenth century men glorified
machines not simply as functional objects but as signs and symbols
of the future of America. They beheld plumes of steam and whirling wheels with a double vision, like a father over his infant son,
inflating each movement with dreams of what it might presage.[61]

One sees this generalizing urge most clearly in a utopian tract
of the period, John Adolphus Etzler's *The Paradise within the
Reach of all Men, without Labor, by Powers of Nature and*

Machinery (1833). Etzler, a German émigré and engineer who led a group of his countrymen to resettle in America, was a technological and social visionary: a man consumed by the possibilities of harnessing the forces of nature—sun, waves, wind, and tides—for the service of mankind. In his book he promised "to show the means for creating a paradise within ten years, where every thing desirable for human life may be had for every man in superabundance, without labor, without pay; where the whole face of nature is changed into the most beautiful form of which it be capable; where man may live in the most magnificent palaces, in all imaginable refinements of luxury." Here was prosperity indeed! In Etzler's vision luxury was no longer a temptation to be shunned but a blessing to enjoy, as machines labored tirelessly to create a garden of earthly delights. Oliver Evans's earlier plan for a totally mechanized mill had been expanded in Etzler's ecstatic conception to cover the globe. Yet in contrast to the painstaking drawings with which Evans explained his design, Etzler provided no illustrations whatsoever. Though he referred vaguely to the addition and substitution of various cranks to achieve his aim, Etzler displayed sublime indifference to any request for particulars. "Machineries are but tools," he replied. "The possibility of contriving tools for any certain purpose cannot be questioned." Evans was an inventor but Etzler was a prophet; he devoted himself to dreams of technological perfection, confident that with modern machinery nothing was impossible.[62]

Etzler's specific vision was certainly eccentric, but not his broader faith in the transforming power of American technology. By the second quarter of the nineteenth century, defenders of the social progress afforded by technology had indisputably gained the offensive over their critics, and they harnessed the destinies of technology and republicanism more tightly than ever before. Such celebrants were by no means confined to any specific political party, region, or class; enthusiasm for republican technology infused the culture and pervaded public discussion. Nevertheless, some of the purest examples and most ringing affirmations of the emergence of republican technology in the ante-bellum period are to be found in the speeches of the leading Whig orators of the Northeast who articulated the social vision behind the dramatic expansion of American industry, and particularly in the works of Edward Everett.

The leading political orator of the day after Daniel Webster, Everett achieved through his powers of eloquence a distinguished and varied career. He amassed a brilliant record at Harvard College, earned an advanced degree in divinity, and was appointed minister of Boston's prestigious Brattle Street Church (Unitarian), all before his twentieth birthday. Slightly over a year later, he was chosen to fill a newly endowed chair in Greek literature at Harvard. In preparation for the position, Everett spent almost five years of diligent travel and study in Europe, becoming in 1817 the first American to receive the Ph.D. from Göttingen.[63] Upon his return, he supplemented his duties at Harvard as editor of the distinguished journal the *North American Review*. The gifts that distinguished Everett as preacher, teacher, and writer proved a springboard to politics. Beginning as a member of Congress, he served in turn as governor of Massachusetts, minister to the Court of St. James, and, after an interlude as president of Harvard, secretary of state under Millard Fillmore, and finally, senator from Massachusetts. Upon his death in 1865, Richard Henry Dana hailed him as a speaker and writer whose fame has been "fairly earned and is firmly fixed."[64]

Despite Dana's eulogy, as the popularity of oratory has declined, Everett's reputation has been eclipsed. Yet if Everett's eloquence appears hollow to the modern reader, he is nonetheless valuable to the historian. In the ante-bellum period, oratory was revered as the queen of the arts and the orator as a figure of potentially inspired vision, capable of sweeping his audience together with a sense of common values and purpose and of articulating their deepest beliefs.[65] Upon hearing Webster's oration at Plymouth Rock in 1820, the cultivated scholar George Ticknor reported, "I was never so excited by public speaking before in my life. Three or four times I thought my temples would burst with the gush of blood."[66] Though Everett lacked Webster's power, he was renowned for the elegance and beauty of his addresses. Such was his influence that Ralph Waldo Emerson, while critical of Everett's limitations, treasured throughout his life a vivid recollection of his first exposure to Everett's lectures and sermons while an undergraduate at Harvard College: "There was an influence on the young people from the genius of Everett which was almost comparable to that of Pericles in Athens.... If any of my readers were at that period in Boston or Cambridge, they will easily remember his radiant beauty of person, of a classic style, his heavy large eye, marble lids, which gave the impression of mass which the slightness of his form needed; sculptured lips; a voice of such rich tones, such precise and perfect utterance, that, although slightly nasal, it was the most mellow and beautiful and correct of all the instruments of the time." Because Everett sought and achieved such influence and rapport with his audience, his speeches offer an important perspective upon a significant segment of public opinion. As Perry Miller has slyly remarked, "No other orator so elegantly presented platitudes to the populace, because no one else so fervently believed them to be exertions of the brain."[67]

Everett's speeches mark the culmination of the alliance between American republicanism and modern technology initiated during the Revolutionary period. The expansive note of new possibilities afforded by republican technology, which late eighteenth-century Americans had sounded, swelled in the orations of Everett and his contemporaries into a full-throated paean to progress. "We live in an age of improvement...," Everett exclaimed; "what changes have not been already wrought in the condition of society! what addition has not been made to the wealth of nations and the means of private comfort by the inventions, discoveries, and improvements of the last hundred years!" All nations, of course, stood to gain from these improvements, but according to Everett, the blessings of nature and republicanism placed Americans in a uniquely advantageous position. Unhampered by governmental or social restrictions, both labor and capital might seize the opportunities of an expanding nation and leap to the challenges of the new technology. While society was "full" and opportunities limited in Europe, the effort of settling the American continent was just beginning. To this task technology would provide abundant assistance, in effect opening up a whole new frontier for the energetic nation:

> The older parts of the country, which have been settled by the husbandman, and reclaimed from the state of nature, are now to be settled again by the manufacturer, the engineer, and the mechanic. First settled by a civilized, they are now to be settled by a dense, population. Settled by the hard labor of the human hands, they are now to be settled by the labor-saving arts, by machinery, by the steam engine, and by internal improvements.

All the while, western lands would protect America from the evils of overpopulation suffered abroad. Harnessing technology to his very rhetoric, Everett hailed cheap and plentiful land as "a safety valve to the great social steam engine," thereby popularizing a metaphor that would justify American industrial expansion through most of the nineteenth century. He did not worry, as would Frederick Jackson Turner sixty years later, about the potential social explosion when all the land was at last settled and the safety valve shut off. For in fact, Everett could not imagine an end to American technological growth. He acknowledged the tendency in the wake of startling recent discoveries and inventions, "to think a pause must follow; that the goal must be at hand." Yet, Everett averred, "There is no goal; and there can be no pause; for art and science are, in themselves, progressive and infinite. . . . Nothing can arrest them which does not plunge the entire order of society into barbarism."[68] The trajectory of America's prosperous advance, as Everett described it, was not cyclical, but ever outward. Declension and disaster would occur only if someone attempted to apply the brakes.

The choice that Everett presented between technological progress and barbaric degeneration was a characteristic one among proponents of technology. The insistence with which it was made reveals how firmly technology had been embraced by American civilization in the second quarter of the nineteenth century and absorbed into an ethic of progress. In this spirit Harvard professor Jacob Bigelow, who helped to introduce the term "technology" in America, proclaimed it the basis and distinction of modern civilization. As knowledge increased from generation to generation, Bigelow argued in his book *Elements of Technology* (1829), nineteenth-century man had achieved physical power, bodily freedom, and earthly dominion of which the ancients could only dream in fables. History thus emerged as a record of continuous improvement, so that from Bigelow's vantage point, even the period of the Revolution appeared more a bronze than a golden age: "The augmented means of public comfort and of individual luxury, the expense abridged and the labor superseded, have been such, that we could not return to the state of knowledge which existed even fifty or sixty years ago, without suffering both intellectual and physical degradation."[69] In a book published for circulation in Massachusetts' public schools, Alonzo Potter, later Episcopal bishop of Pennsylvania, cast the issue in even starker terms. Naturally "one of the most defenceless and wretched" of all creatures, man by virtue of technology effortlessly ruled the earth:

> Not only animals, with their fleetness and strength, but even winds, and waves, and heat, and gravity, have been trained to obey him; and, operating by means of machinery, they now fabricate for him, almost without intervention on his part, the choicest food and raiment; transport him, with the celerity of the deer or the antelope, from place to place; and surround him with all the comforts and conveniences of life.[70]

What Everett did was to endow these arguments with his own weightier diction and give them a characteristically moral emphasis. The useful arts, he declared in a speech of 1831, constituted both the cause and product of civilization and formed "the difference between the savage of the woods and civilized, cultivated, moral, and religious man." Everett conceived of knowledge as increasing by the collective efforts of mankind over the course of history, and saw in technology the development of human civilization in microcosm: the wisdom of the ages and the underpinning of both society and religion. If Lyman Beecher insisted upon the importance of Christianity in safeguarding American technology, Everett replied that the converse was true. Thus, he exhorted his audience to faith and diligence in accents befitting an industrial age. Technology stood as the great benefactor of the public good, and he who impeded the progress of modern inventions threatened all: "The moral and social improvement of our race, and the possession of the skill and knowledge embodied in them, will advance, stand still, and fall together."[71]

Viewed from such a perspective, modern machinery became for Everett and his contemporaries manifestations of the sublime, achievements of mind that challenged the powers of comprehension and description. "There is an untold, probably an unimagined, amount of human talent, of high mental power, locked up among the wheels and springs of the machinist; a force of intellect of the loftiest character," Everett exclaimed on another occasion. "This stunning din, this monotonous rattle, this tremendous power, and the quiet, steady force of these humble, useful, familiar arts, resulted from efforts of mind kindred with those which have charmed or instructed the world with the richest strains of poetry, eloquence, and philosophy." Grasping for a simile to express his sense of wonder, Everett, like so many writers of his age, hailed the inventor as a modern magician able to manipulate the world about him at a whim: "He kindles the fires of his steam engine, and the rivers, the lakes, the ocean, are covered with flying vessels. . . . He stamps his foot, and a hundred thousand men start into being; not, like those which sprang from the fabled dragon's teeth, armed with the weapons of destruction, but furnished with every implement for the service and comfort of man."[72]

Everett thus exulted in the gigantic strides of American progress through technology. His confidence was complete. Nowhere in his writings is there any suggestion that American technological development need in any way be directed or controlled; only let it be encouraged by every possible means, it could not help but result in ever higher stages of civilization. The few, fleeting references in his speeches to the possibility of deterioration and social collapse were directed at those who sought to restrict American industrial development. To critics of industrial capitalism, for example, Everett responded with incredulity: "What is your object? . . . Do you wish to lay on aching human shoulders the burdens which are so lightly born by these patient metallic giants?" Cities such as Lowell, converted by capital from a few poor farms into a thriving and "noble city of the arts," resembled a miracle out of *The Arabian Nights*, and those who carped at such achievements were playing "the part of the malignant sorcerer, in the same Eastern tale, who potent only for mischief, utters the baneful spell which breaks the charm, heaves the mighty pillars of the palace from their foundation, converts the fruitful gardens back to their native sterility, and heaps the abodes of life and happiness with silent and desolate ruins." The blame for declension, if it came, would lie not upon modern technology, its factories, and prosperity, but on their enemies.[73]

Even these fears, of course, were more rhetorical than real. Certain doubts, it is true, occasionally surface in the rhetoric of even the most passionate admirers of technology. Monstrous or demonic imagery affects their descriptions of machines they profess to admire; ostensibly cheerful concessions of the simplicities of the past and the inconveniences and dangers of modern life reveal a sense of anxiety in the course of events.[74] Important as it is to be aware of these misgivings, they are easily overemphasized. In the case of so many Americans, what is truly striking is the resiliency of their abounding faith and delight in technology, even when it conflicted with their own experience. Take the extraordinary popularity of steamboats and railroads as an example. Daniel Boorstin has estimated that of all steamboats built before 1850, roughly 30 per cent were lost in accidents, and on western rivers the figures ran much higher. A convocation of steamboatmen which met in 1838 to consider the causes of such accidents concluded that the "morbid appetite among travelers for 'going ahead' is probably one of the greatest causes of the evils." The situation was much the same with railroads, and European visitors were frequently shocked by dangerous conditions that Americans accepted as routine. Riding on an American train whose conductor was determined to make up time could be a harrowing experience, as Charles Richard Weld, an English tourist, discovered in 1855. As his train sped along, Weld's car began shaking violently and its ceiling lamps smashed to the floor. He remonstrated to the conductor, but to no avail. When finally the train jumped the tracks, all the cars except half of Weld's own and the engine were smashed. Yet to his amazement none of his fellow passengers wished to join him in protest; most, on the contrary, applauded the conductor's efforts to arrive on time.[75]

In its stark depiction of travelers so consumed by their vaunted technological improvements that they refused to recognize near-fatal disaster even as they experienced it, Weld's anecdote approaches fable. Indeed, it recalls one of the sharpest sketches of how far Americans had departed from the world of their ancestors in their pursuit of technological and social progress, Nathaniel Hawthorne's brilliant satire, "The Celestial Railroad" (1843). Here Hawthorne offered a revision of John Bunyan's *Pilgrim's Progress* for a technological age. Visiting the region of Christian's historic pilgrimage, Hawthorne's narrator discovers that only old-fashioned pilgrims continue to toil on foot from the City of Destruction to the Celestial City. For more modern travelers, a railroad has been established which eases the journey considerably, serviced by Prince Beelzebub's subjects, with Apollyon presiding as chief engineer. Guided by a director and leading stockholder in the celestial railroad, Mr. Smooth-it-away, the narrator joins a large and prosperous group of passengers. They proceed merrily, passing by modern tunnel through the Hill of Difficulty and speeding through the Valley of the Shadow of Death, their way brightened by gas lamps. After a long sojourn in the gay and prosperous city of Vanity Fair, now a thriving center of liberal religion, the narrator resumes his modern pilgrimage, still accompanied by Mr. Smooth-it-away. At last they near the Celestial City, only to see the two poor pilgrims on foot who had been taunted by Apollyon and the passengers at the commencement of their journey arriving at the celestial gates. Suddenly anxious and despairing, the railroad passengers scramble in panic onto a sinister steam ferryboat for the last leg of their trip. The narrator turns to discover Mr. Smooth-it-away waving good-bye and—with a diabolical laugh, a snort of smoke, and twinkle of flame—promising that they will meet again. Thus, the road to hell is paved with good inventions. The celestial railroad is ultimately an infernal one.[76]

In "The Celestial Railroad" Hawthorne offered a burlesque of the progressive vision of both nineteenth-century religion and technology and the prosperous and complacent moral climate in which the two flourished. The apostles of religious and technological perfectionism, as viewed in Hawthorne's story, attempted to provide shortcuts to salvation, to assure their prosperous followers that moral and social progress could be achieved through external contrivances instead of individual virtue and effort. Here, as throughout his work, Hawthorne remained profoundly skeptical of the pretensions of nineteenth-century American "improvements" or of any attempts to apply panaceas to the problems of society and the human heart.[77]

But one satirizes only what is established, sanctified. In dissenting from the progressive spirit of the age, Hawthorne implicitly testified to its power and authority. His skeptical gaze was a minority view. Everett's ebullient vision of technological and social progress was most keenly attuned to the nation's. Like Mr. Smooth-it-away, Everett affirmed that mechanical improvements had revolutionized the once arduous republican journey. On his celestial railroad Americans would travel smoothly and commodiously toward an ever more glorious future. Both the words and tone of his orations denied any sense of contradiction, reservations, or painful adjustments that might have to be made. The inherent complexities that he blithely swept aside were left to other commentators and events to uncover.

The connections between technology and American culture from the Revolutionary period to the Jacksonian era were complex. It would be far too simple to reduce them to formulas, either that "ideology follows technology" or that "technology follows ideology."[78] Their relationship was rather a dialectical interchange in which both were transformed. Republicanism developed into a dynamic ideology consonant with rapid technological innovation and expansion. The older moral imperatives of eighteenth-century republicanism were modified to suit a new age of industrial capitalism. As technological progress offered new stability for republican institutions, luxury lost its taint. For its part, the ideology of republicanism helped to provide a receptive climate for technological adaptation and innovation. The promise of labor-saving devices strongly appealed to a nation concerned with establishing economic independence, safeguarding moral purity, and promoting industry and thrift among her people. So too did the hope that increased production, improved transportation and communications would centralize a country that continued to fear regional fragmentation. Yet the union of technology and republicanism, while settling some issues, raised others. Particularly pressing was the question whether the new centers of American production, her manufacturing towns, could avoid the blight and degradation of their English counterparts and achieve a new standard as model republican communities. If not, then Jefferson's worst fears might stand confirmed after all.

2

The Factory as
Republican
Community

LOWELL, MASSACHUSETTS

THE QUESTION of what social environment American manufactures would create went to the heart of the republican venture. The introduction of new manufacturing centers portended dramatic changes in the structure of society. Their impact upon the character of American life was an issue of national concern. Could a system of manufactures be established that would nurture and protect the health, intelligence, independence, and virtue of their operatives, qualities essential to a republic? Or would factories breed disease, ignorance, dependence, and corruption? Would industrialization provide new prosperity and comfort for all levels of society? Or would industrialization prove an instrument of economic and political repression and social cleavage? In short, was the Revolutionary ideal of a republican civilization compatible with rapid industrial development? On the answer to these questions much of the nation's future depended.

Americans in the early nineteenth century united in admiration of English machine technology; smuggling British industrial secrets and mechanics was the sincerest form of flattery. However, there was considerably less enthusiasm for the social consequences of the English factory system. Jefferson found cause to revise his earlier opposition to the promotion of domestic manufactures, but not his horror of the "mobs" of workmen in European cities. In the late eighteenth and early nineteenth centuries, factory towns sprang up in England at unprecedented rates, stimulated by the colossal expansion in cotton manufactures in Lancashire. The capital of the cotton industry, Manchester, expanded from an ancient town of 17,000 people in 1770 to over 70,000 by 1801, 142,000 in 1831, and over 250,000 by midcentury, with more than an additional 150,000 in the sprawling towns that surrounded it. It stood as the "shock city" of the age, attracting numerous visitors both from England and abroad anxious to confront the symbol and embodiment of the new industrial order. Manchester's contrasts both fascinated and repelled: the advanced technology and immense productivity of its factories; the unbelievably primitive, cramped, and diseased hovels; the vitality of its magnates; the feebleness and despair of its workers. Wrestling with its conflicting characteristics during a visit in 1835, the astute social critic Alexis de Tocqueville concluded: "From this foul drain the greatest stream of human industry flows out to fertilise the whole world. From this filthy sewer pure gold flows. Here humanity attains its most complete development and its most brutish; here civilisation works its miracles, and civilised man is turned back almost into a savage."[1]

In their effort to probe the significance of Manchester as a new social phenomenon, observers again and again returned to its dramatic if horrific physical presence. As early as 1808 a visitor reported, "The town is abominably filthy, the Steam Engine is pestiferous, the Dyehouses noisesome [sic] and offensive, and the water of the river as black as ink or the Stygian lake." As the city grew, this image of Manchester as an industrial inferno was amplified and expanded. Tocqueville, who also called the polluted river "the Styx of this new Hades," depicted a city that assaulted all one's senses:

> Heaps of dung, rubble from buildings, putrid, stagnant pools are found here and there among the houses and over the bumpy, pitted surfaces of the public places.... Look up and all around this place you will see the huge palaces of industry. You will hear the noise of furnaces, the whistle of steam. These vast structures keep air and light out of the human habitations which they dominate.... A sort of black smoke covers the city. The sun seen through it is a disc without rays. Under this half daylight 300,000 human beings are ceaselessly at work. A thousand noises disturb this damp, dark labyrinth, but they are not at all the ordinary sounds one hears in great cities.
>
> The footsteps of a *busy* crowd, the crunching wheels of machinery, the shriek of steam from boilers, the regular beat of the looms, the heavy rumble of carts, these are the noises from which you can never escape in the sombre half-light of these streets.[2]

The vividness, detail, and urgency of Tocqueville's description recurred in personal accounts, official reports, and industrial novels in the 1830s, '40s, and '50s as writers of widely varying backgrounds and beliefs converged on a common strategy of social documentary. Describing the "facts" of the city, they chose those facts most expressive and often shaped them for greatest emotional impact in the effort to comprehend and convey a situation at once "immediate and all but unimaginable."[3] Thus the same ingredients of Tocqueville's 1835 account—the Stygian river, looming factories, labyrinthine streets, dark and putrid working quarters—were presented in even greater abundance in Friedrich Engels's study of Manchester, *The Condition of the Working Class in England in 1844*; tinged with metaphor, they reappeared a decade later in Dickens's fictional portrayal of Coketown in *Hard Times*:

> It was a town of red brick, or of brick that would have been red if the smoke and ashes had allowed it; but, as matters stood it was a town of unnatural red and black like the painted face of a savage. It was a town of machinery and tall chimneys, out of which interminable serpents of smoke trailed themselves for ever and ever, and never got uncoiled. It had a black canal in it, and a river that ran purple with ill-smelling dye, and vast piles of buildings full of windows where there was a rattling and a trembling all day long, and where the piston of the steam-engine worked monotonously up and down, like the head of an elephant in a state of melancholy madness.[4]

Such a city challenged traditional notions of class, community, the entire social order. Commentators had observed a "growing gulf" between rich and poor as early as the 1780s, and in the first half of the nineteenth century, particularly after the Peterloo massacre of 1819, the gulf appeared to have become a chasm creating, in Disraeli's phrase, "two nations." The result, according to one English clergyman who visited Manchester, was a new industrial feudalism, more stratified and severe than any England had ever seen: "There is far less *personal* communication between the master cotton spinner and his workmen, between the calico printer and his blue-handed boys, between the master tailor and his apprentices, than there is between the Duke of Wellington and the humblest labourer on his estate, or than there was between good old George the Third and the meanest errand-boy about his palace."[5]

For American travelers—less affectionately disposed toward George III and acutely sensitive to the problem of social cohesion—the encounter with Manchester often proved especially disturbing. While the capacity and ingenuity of English technology frequently compelled their admiration, they were shocked by the extremes of rich and poor, of "princely wealth and abject poverty, of lordly power and cringing servility." The gross inequities of social condition they witnessed substantiated their opinion of the limitations—even the corruption—of English institutions, and they toured the streets with a sense of impending revolution. The spectacle of industrial conditions in Manchester and other English cities became an "emotionally validating image" confirming their belief in the sanctity of American republicanism, located within a predominantly agrarian setting. Although not opposed to manufacturing per se, they resolved that never should the United States develop a factory system like that which they confronted in England.[6]

One sees this pattern of response occurring as early as 1805, when the young Yale professor Benjamin Silliman, one of the most eminent American scientists of his generation, visited Manchester in the course of his scientific travels abroad. The technology of Manchester's great factories, "the wonder of the world, and the pride of England," greatly impressed him, and while apologizing for what little he could contribute to the subject, he filled his journal with pages of detailed description of various processes. However, Silliman was troubled and repelled by the condition of the operatives. Working in hot, cramped, cotton-choked rooms and leading reportedly debauched lives, they appeared to him "at best, but an imbecile people." From this dismal scene Silliman emotionally recoiled to a pastoral image of American "fields and forests, in which pure air, unconstrained motions, salubrious exhalations, and simple manners, give vigour to the limbs, and a healthful aspect to the face." Nevertheless, he acknowledged that manufactures were essential, even while regretting "the physical, and . . . moral evils which they produce."[7]

The infernal factory city: Manchester, England.

By the 1820s travelers' reports of the debased conditions of English factory workers had grown so common that the New York author James Kirke Paulding could don the persona of an American traveler in England and pen an energetic denunciation of the operatives' oppressed state without ever actually venturing abroad at all. Stung by British criticisms of the United States, in *A Sketch of Old England* (1822) Paulding fervently defended the superiority of America's republican civilization by examining Britain herself. The exploited position of workers in Manchester, Birmingham, and other industrial cities he singled out for special comment. "No one, that has not seen," he wrote, though of course Paulding himself had not, "can conceive the squalid and miserable looks of these people, between the dirt and unwholsomeness [*sic*] of their employment, the ignorant worthlessness of their characters, and the shifts the poor creatures are obliged to resort to in order to exist." Paulding did not wonder that, trapped in such misery, Britain's workers grew rebellious and violent. He affirmed their right to demand a proper share of happiness and prosperity from their government, or else to seek redress, in the words of the Declaration of Independence, "peaceably if they can; forcibly if they must." Although Paulding did not realize it, he pronounced a standard of republicanism that would later be applied to American factories as well.[8]

Other American travelers confirmed Paulding's armchair impressions that the English manufacturing system was fundamentally antirepublican. A superintendent of one of New York State's largest cotton mills, touring English factories in 1840, was challenged by a British colleague: "How do you manage to get along with republican operatives? *I* never would superintend a factory where I could not do as I pleased with my hands. Here we can *make them behave*; they know they are in our power, where they ought to be, and they *walk straight*. . . . I have been in the United States, and I wouldn't stay there. You can't find a man, woman, or child there, that don't feel as good as his employer." C. Edwards Lester, who visited Manchester the same year, returned appalled by the poverty and misery he observed. In the face of such slavery, which American operatives with their republican traditions would never tolerate, he pronounced the English people "ripe for revolution." The horror of England's industrial oppression redoubled Lester's affection and concern for his native land, and he declared: "Heaven forbid that America should ever be cursed with such a manufacturing system as that which is now the curse of England. May the day never come, when any great proportion of the labouring classes of America shall be taken from her broad fields and rich soil, where the muscles grow strong and the frame sturdy by honest labour in the open air; where the wages of a few months will purchase the fee-simple forever of enough of the earth's surface to be dignified by the name of *home*, and which will produce the grand necessaries of life for the working man's family."[9]

To such a prayer, other travelers said "amen." The Unitarian minister and agricultural writer Henry Colman visited Manchester in 1843 and was led on a midnight tour of its working-class quarters by a government sanitary commissioner, accompanied by two policemen. He discovered "exhibitions of the most disgusting and loathsome forms of destitution, and utter vice and profligacy," and emerged from the excursion "shocked . . . with horror." He could not describe the details of his adventure, Colman wrote a friend: "The paper would, I fear, be absolutely offensive to the touch." The experience, he maintained, "will make my life hereafter an incessant thanksgiving that my children have not in the inscrutable dispensation of Heaven been cast destitute, helpless, and orphans in such a country as this." Exposure to industrial conditions in England led ultimately to a renewed sense of the preciousness of American republicanism.[10]

Such reports confirmed the popular American image of English factory towns in the first half of the nineteenth century as centers of advanced technology and productivity but also as cancers against both nature and society, producing an oppressed, ignorant, and debauched working class and threatening the civilization as a whole. Could the United States develop a system of manufactures that would avoid a similar fate? If American technology could indeed, as its proponents from Coxe to Everett claimed, integrate the country socially and politically and buttress its republican virtue, it would have to prove it first at the local level in the nation's new manufacturing towns. Here more than anywhere else would be the testing ground of the new republican industrial order.

No one was more aware of this challenge than American manufacturers themselves. The merchant-entrepreneurs who created the leading industrial towns of the nineteenth century shared their fathers' sense of republican mission and distrust of aristocratic Europe. Though some advocates of manufactures took heart in reports that pauperism pervaded England's agricultural counties to a much greater extent than her manufacturing ones,[11] they were not generally inclined to dispute the sordid reputation of English factory towns. Many of them had observed firsthand what Nathan Appleton called the "misery and poverty" of English industrial workers, and they resolved that American manufactures must never be allowed to take a similar course. Manufacturing itself need not be debilitating, they reasoned. Many of the social and moral evils of the English system, they believed, stemmed from the establishment of factories in large cities, in which vice thrived unchecked and a debased proletariat perpetuated itself. They shared the faith of some of the earliest American planners of industrial towns, including Tench Coxe and Alexander Hamilton, that by locating American manufactures in the countryside and instituting a strict system of moral supervision, the health and virtue of operatives would be protected. Thus situated, manufactures would harmoniously complement agricultural life, and the nation's agrarian character would remain undisturbed.[12]

However, the leading American factory towns of the first half of the nineteenth century were shaped not only in response to the English factory system but to events in America as well. As we have seen, technology was absorbed into a conservative ideology of republicanism early as the 1780s in part as an instrument of social order and control against both the insidious influences of European manufactures and symptoms of social discord and rebellion at home. As Americans advanced into the nineteenth century, pressures on a deferential society continued and the problems of republican order increased. The whole country surged with dramatic volatility and energy. The nation's population, which had more than doubled every twenty-five years in the eighteenth century, continued to grow at the same phenomenal rate through the first half of the nineteenth. People migrated restlessly not only along the vast new frontier but within the rapidly mushrooming urban centers as well.[13] And the concept of republicanism, instead of controlling and containing this expansion, became in the hands of new egalitarian forces a weapon with which to challenge established authority in politics, religion, law, commerce—virtually every aspect of society. Social conservatives rubbed their eyes to see a reversion in American life from civilization to barbarism as the whole social order upon which the republican experiment was premised appeared to be collapsing around them. Some recent scholars, including Stanley Elkins and David Donald, have in effect supported their perception, arguing that ante-bellum America suffered from a general "institutional breakdown" and "an excess of democracy" which ultimately paved the way for Civil War.[14]

But the ante-bellum period was a time of institution building as well as breaking. Countervailing an intense anti-institutionalism, widespread attempts were made to create *new* institutions capable of dealing with an increasingly complex urban and industrial society and to restore social cohesion and public virtue. The colonial model of a relatively stable, hierarchical community in which each person observed established codes of morality and deference still exerted a powerful influence and appeal over social leaders disturbed about the new direction American society seemed to be taking. Yet at the same time they also wished to grasp the expansive new opportunities which beckoned. They felt with particular intensity the culture's ambivalence toward urbanism and industrialism. In this vision, cities and factories were seen as necessary and laudable agents of prosperity; but they were also viewed as potential tinderboxes of corruption and mob disorder. The promise of increased progress and abundance was far too enticing to allow nostalgia to sweep any significant segment of American society into a reactionary mood. Instead of attempting to solve their dilemma by simply turning back the clock, then, social conservatives concentrated on devising new methods of social improvement and control. They sought to regulate the fluid urban industrial society so as to safeguard their vision of an ordered republic while at the same time enjoying its benefits. In this effort they endeavored to create and harness institutions of moral instruction, rehabilitation, and reform. In the past few years studies by a number of historians have documented the range and extent of their efforts: in public education; in support of the fine arts; in Protestant missions; in the treatment of poverty, delinquency, crime, and insanity.[15] Public schools, parks and gardens, art galleries and museums, Sunday schools and tract societies all represented attempts to extend the sphere of republican instruction in the principles of social order and virtue to the maximum number of citizens; to counteract the turbulence and corruption of American life by improving the social environment and establishing monitors over it. In cases where the existing social conditions were potentially most dangerous or the subject most in need of discipline, the commitment to institutional reform was most complete. Thus in the decades after 1820, as David Rothman has shown in his book

The Discovery of the Asylum, the almshouse, orphanage, penitentiary, reformatory, and insane asylum all were erected and meticulously systematized to deal with the deviant and the dependent.
Often established in country settings, these various asylums
sequestered inmates from outside influences and organized their
lives around a routine of disciplined and officially sanctioned
conduct. The order, supervision, industry, and temperance of
institutional life, authorities believed, would counteract the chaotic
and corrupting forces of the larger society and transform social
victims into respectable citizens. More generally, such asylums
would stand as models of well-ordered institutions for a society
desperately in need of standards and guidance. They were part of
an increased reliance in the second quarter of the nineteenth
century with what, in Erving Goffman's illuminating concept, we
may call "total institutions"—that is, places "of residence and
work where a large number of like-situated individuals, cut off
from the wider society for an appreciable period of time, together
lead an enclosed formally administered round of life." The total
institution represented the ultimate form in this period of the
search for institutions of republican community.[16]

To the total institution, then, turned a group of merchants
known as the Boston associates, who would become America's
leading manufacturers before the Civil War, as they sought an
alternative to the poverty and neglect of English industrial conditions and a safeguard against the fluidity and potential corruption of an expanding American society. Beginning in Waltham,
Massachusetts, in 1815, they established a successful pattern of
textile manufactures and extended it rapidly. By 1850 the Boston
associates controlled mills in operation in Chicopee, Taunton, and
Lawrence, Massachusetts; Manchester, Dover, Somersworth, and
Nashua, New Hampshire; and Saco and Biddeford, Maine; and
were making active preparations for new mills in Holyoke,
Massachusetts.[17] But the queen city of their system and the leading
producer of cotton goods, the nation's largest industry before the
Civil War, was Lowell, Massachusetts. Lowell's fame rested not
only on its industrial capacity but even more on its reputed social
achievement. One of the most important and influential of all total
institutions of republican reform in the ante-bellum period, Lowell

The celestial factory town: Lowell, Massachusetts.

promised to resolve the social conflict between the desire for industrial progress and the fear of a debased and disorderly proletariat. Its founding sprang from the conviction that, given the proper institutional environment, a factory town need not be a byword for vice and poverty, but might stand as a model of enlightened republican community in a restless and dynamic nation. Lowell offers a dramatic example of the effort to put this conservative faith into practice. Its story is particularly interesting because within a few years of its founding, the basic assumptions of the Lowell factory system and its conception of republican community were challenged both on ideological and institutional grounds by the working class and their spokesmen. Branding Lowell's directors as a repressive new aristocracy, dissident workers increasingly rejected what they regarded as a manipulative social structure and an exploitative industrial capitalism. Against the conservative view of republicanism of Lowell's directors, pro-

testing workers interpreted the American Revolution as the beginning of a continuing struggle toward a radical egalitarianism.
The early history of Lowell thus provides an encapsulated version
of the debate over the meaning of republicanism in an industrial
society and the attempt to give that meaning institutional shape.[18]

Lowell was conceived in the second decade of the nineteenth
century by a trio of innovative and energetic young Boston
merchants: Francis Cabot Lowell, Nathan Appleton, and Patrick
Tracy Jackson. Touring Great Britain in 1810 and 1811 for his
health, F. C. Lowell visited a large iron works in Edinburgh and
grew excited over the enormous possibilities such large-scale
manufacturing had for America. While in Edinburgh, he also met
Nathan Appleton, his friend and fourth cousin, and the two
merchants discussed the idea of establishing cotton manufacture
employing English technology in the United States.[19] At the same
time Lowell was corresponding on the subject with his business
partner and brother-in-law, P. T. Jackson, and he determined, before his return to America, to study thoroughly the cotton mills
at Manchester and Birmingham. He spent weeks in these factories,
applying his keen mathematical and mechanical skill and questioning engineers eager to accommodate a wealthy potential customer.
Thus Lowell circumvented stringent regulations against the exportation of English machinery or mechanical drawings and smuggled into America valuable mental baggage. His contemporaries
would later acclaim him a hero and a genius, who had performed
an act of patriotic espionage to rank with Samuel Slater's a generation earlier.

Shortly after Lowell's return from Europe, he and Jackson
bought a water-power site in Waltham, obtained a charter of incorporation from the Massachusetts legislature for their new
Boston Manufacturing Company, and sought investors for the
enterprise within their circle of friends and relatives among
Boston's merchants. Some of Lowell's relations, including the
Cabots whose pioneering 1787 cotton factory at Beverly had
failed, attempted to dissuade him from what they considered "a
visionary and dangerous scheme, and thought him mad." Nathan
Appleton himself warily agreed to invest only five thousand

dollars, half the amount Lowell and Jackson requested, "in order to see the experiment fairly tried." The two merchants also enlisted the financial support of Patrick Jackson's brothers; Israel Thorndike and his son; Uriah Cotting; James Lloyd; and two of Lowell's brothers-in-law, Benjamin Gorham and Warren Dutton.[20]

Lowell hired a talented engineer, Paul Moody, and quickly set about a series of reinventions based upon his observations of English machinery and contemporary American developments. Of these the most important was the power loom, which promised to free American mills from dependence on neighborhood weavers and to permit the organization of all manufacturing processes from raw cotton to finished cloth within a single integrated mill complex. When Nathan Appleton first saw Lowell's loom in 1814, he was stupefied by its significance and, in a "state of admiration and satisfaction," sat with Lowell "by the hour, watching the beautiful movement of this new and wonderful machine, destined as it evidently was, to change the character of all textile industry." To exploit the capacity of large-scale mechanized production to its fullest extent while relying on unskilled labor, Lowell decided to concentrate production on standardized inexpensive cotton cloths, sheetings, and shirtings. Later, as new corporations arose at the town of Lowell and elsewhere, each manufactured a different type of cotton goods to avoid duplication and competition with fellow companies. Mills were designed to facilitate the flow of materials from one stage of processing to the next. Cotton was carded on the first floor, spun on the second, woven on the third and fourth, while machine shops resided in the basement. In the next fifteen years New England inventors would build upon this structure and introduce a series of labor-saving technological innovations which equaled or excelled British methods and machinery and mechanized all the basic processes of cloth manufacturing except spooling and warping. Even before some of these refinements, however, Lowell's system achieved dramatic gains in production. According to one technological historian, from its first years of operation the Waltham mill could with the same number of employees produce three and a half times as much as other American factories still operating according to pre-1812 methods. The achievement of Lowell and his colleagues, sometimes known as the "Massachusetts system," thus marked a significant stage in the development of modern mass production.[21]

As a final stroke in his grand design, Lowell turned his attention to politics. Competition with British textiles had in the past been the bane of the American industry. Thus when Congress began deliberations over a new tariff measure in 1816, Lowell rushed to Washington to lobby for his cause. He adroitly steered through Congress a minimum valuation tariff which helped to establish the principle of protection to American industry and sheltered his own company's products from foreign competition, while leaving exposed rival manufacturers of more expensive cotton goods. Lowell made a powerful impression even on opponents of the protective tariff, such as Daniel Webster, then a representative from New Hampshire. Only two years earlier, discussing another tariff measure Webster had declared he was "not in haste to see Sheffields and Birminghams in America." The grim image of English industrial towns dominated his thinking on the subject, and he gestured with foreboding toward the day "when the young men of the country shall be obliged to shut their eyes upon external nature, upon the heavens and the earth, and immerse themselves in close and unwholesome workshops; when they shall be obliged to shut their ears to the bleating of their own flocks, upon their own hills, and to the voice of the lark that cheers them at the plough, that they may open them in dust, and smoke, and steam, to the perpetual whirl of spools and spindles, and the grating of rasps and saws." Lowell helped Webster to change his opinion and to convert him gradually to the protectionist position. Webster's ambition was outgrowing New Hampshire, and he soon moved to Boston, where Lowell supplied him with letters of introduction. Such ministrations, including a later offer to obtain stock in the Boston associates' new enterprise at the town of Lowell, ultimately won Webster's services as a major apologist for American industrial interests.[22]

In the eyes of his contemporaries, however, Lowell's greatest achievement lay in neither his technological success, nor his political skill, nor his business acumen. The special reverence with which his name was spoken in the period before the Civil War emerged from the sense that he had conceived a manufacturing system that concerned itself as much with the health, character, and well-being of its operatives as it did with profits. By allegedly protecting the integrity of America's workers, he had in important measure safeguarded the character of the republic itself. From the beginning, Lowell and his associates were mindful of the condition of European workers and particularly concerned to avoid a similar fate here. As Appleton recalled their earnest discussions, "The operatives in the manufacturing cities of Europe, were notoriously of the lowest character, for intelligence and morals. The question therefore arose, and was deeply considered, whether this degradation was the result of the peculiar occupation, or of other and distinct causes. We could not perceive why this peculiar description of labor should vary in its effects upon character from all other occupation."[23]

Their solution was to organize the factory as a total institution, so that the company might exercise exclusive control over the environment. Unlike most English cotton factories of this time, which were powered by steam, American mills depended upon water power; and the necessity to locate the plant near an important rapids further insured that the community would be placed in the country, apart from urban contamination. But where Lowell's plan differed radically from both earlier English and American factory settlements was in his decision to establish a community with a rotating rather than a permanent population; this was central to the conception. Previous American factory settlements had retained the English system of hiring whole families, often including school-aged children. Lowell and his associates opposed the idea of a long-term residential force that might lead to an entrenched proletariat. They planned to hire as their main working force young, single women from the surrounding area for a few years apiece. For a rotating work force such women were an obvious choice. Able-bodied men could be attracted from farming only with difficulty, and their hiring would raise fears that the nation might lose her agrarian character and promote resistance to manufactures. Women, on the other hand,. had traditionally served as spinners and weavers when textiles had been produced in the home, and they constituted an important part of the family economy. However, imports of

European manufactured fabrics were eroding American household industry. At the same time, southern New England farmers were gradually shifting from subsistence to commercial agriculture. By employing young farm women in American factories on a relatively short-term basis, the Lowell system in effect extended and preserved the family economy while at the same time avoiding incorporation into the factory of the family as a whole.[24] Factory work, then, would not become a lifetime vocation or mark of caste, passed on from parent to child in the omnipresent shadow of the mill. Rather it might form an honorable stage in a young woman's maturation, allowing her to supplement her family's income or earn a dowry, before assuming what the founders regarded as "the higher and more appropriate responsibilities of her sex" in a domestic capacity. Her factory experience would be a moral as well as an economic boon, numerous spokesmen for American manufactures maintained, rescuing her from idleness, and vice, pauperism, possibly even confinement in an almshouse or penitentiary. Instead, in the cotton mill, under the watchful eyes of supervisors, she would receive a republican education, imbibing "habits of order, regularity and industry, which lay a broad and deep foundation of public and private future usefulness."[25] During her term at Lowell, the worker would be protected *in loco parentis* by strict corporate supervision, lodged in company boardinghouses kept by upright matrons, and provided compulsory religious services. Such stringent standards of moral scrutiny and company control would serve a treble purpose: to attract young women and overcome the reluctance of their parents, most of them farmers; to provide optimal factory discipline and management control of the operatives; and to maintain an intelligent, honorable, and exemplary republican work force. Though Lowell's founders never regarded their efforts as utopian, they aimed to establish an ideal New England community, which would stand not as a blight but a beacon of republican prosperity and purity upon the American landscape.

Recently, however, some scholars have questioned the extent to which the Lowell system actually stemmed from any grand social vision or solicitude in behalf of the workers. How much choice, they ask, did Lowell's founders really have in developing their vaunted system? According to the economist Howard M. Gitelman, the complexity of the early power-driven machinery employed at Lowell and elsewhere made child labor unfeasible, and thus the economies of a family labor system were not a viable option for the founders. Moreover, he contends, the rural location of Waltham, Lowell, and similar mill towns was dictated mainly by considerations of available water power; company housing then had to be provided in order to staff the mills. Concerned parents and an aroused community, Gitelman speculates, would in any case have insisted upon supervised company housing and a strict system of rules and regulations for the operatives. Economic necessity, not employer magnaminity, so the argument runs, compelled the shape of Lowell.[26]

But to conclude that because the Boston associates were not altruistic reformers, they were therefore simply capitalists following the line of least economic resistance clearly ignores a broad middle ground. A fuller, more satisfactory explanation of the founding of Lowell would recognize *both* commercial and social and ideological motives. For the Boston associates and many of their colleagues were in fact both capitalists and concerned citizens, hard-dealing merchants and public-spirited philanthropists, entrepreneurs and ideologues. Even as they helped to transform New England's economy, they sought to preserve a cohesive social order by adhering tenaciously to a rigorous code of ethics and responsibility. They took seriously their role as republican leaders, and the public turned to them for leadership. The Unitarian reformer Theodore Parker expressed the sense of gratitude of many when he praised the development of manufactures and improvements in transportation as helping to "civilize, educate, and refine men." "These are men," he concluded, "to whom the public owes a debt which no money could pay, for it is a debt of life." Whether it was sufficient payment or not, obviously these manufacturers received a great deal of money for their services. Nevertheless, they insisted both publicly and privately that wealth was not their goal. "My mind has always been devoted to many other things rather than money-making," Nathan Appleton declared toward the end of his life. "Accident, and not effort, has made me a rich man." Amos Lawrence, who with his brother Abbott joined forces with Lowell's investors in 1830, filled his diary and letters with reminders of the stewardship and public trust which wealth entailed. From 1829 through 1852 he personally and meticulously made charitable gifts of $639,000 in cash, as well as clothing, food, books, and other articles. He once wrote a factory agent, "We must make a good thing out of this establishment, unless you ruin us by working on Sundays. Nothing but works of necessity should be done in holy time." Boston's leading merchants generally scorned a narrowly acquisitive view of their role and participated in a wide variety of public affairs. They were active and influential in Federalist and later Whig politics and held important offices on both state and national levels. Their contributions to numerous charities and philanthropies, including hospitals, orphanages, and asylums, as well as libraries, historical societies, schools and colleges, helped to make Boston a center of social and cultural institutions in the nineteenth century. Such enterprises, they believed, were essential to the solidity and progress of society. As Francis Cabot Lowell's son John Lowell declared in establishing a series of public lectures, the Lowell Institute, in 1835, "The prosperity of my native land, New England, which is sterile and unproductive, must depend ... 1st on the moral qualities and 2dly on the intelligence and information of its inhabitants."[27]

Concern with the social consequences of Lowell, Massachusetts, as a tight-knit, carefully regulated republican community, then, was certainly consistent with the values and activities of the founders and their associates in a variety of other fields. Moreover, their philanthropic and industrial pursuits were related both historically and institutionally. Nineteenth-century textile mills were direct descendants of the manufacturing societies formed in various American colonies in the eighteenth century and more distant relatives of the work houses of the seventeenth century. Institutions such as the Boston Society for Encouraging Industry and Employing the Poor, established in 1751 and one of the colonies' most important pre-Revolutionary factories, had, as its name indicates, a dual purpose: not only to stimulate American manufactures but to provide work for the destitute; to encourage industry in both senses of the word, under official supervision.[28] Undoubtedly, with increased mechanization in the textile industry, commercial motives were uppermost in the establishment of Lowell and other mill towns in the nineteenth century, but at the same time one should not lose sight of the social vision that accompanied them. Of course Lowell's founders and directors were not always as idealistic as they professed. But in instituting their factory system, they did not have to choose between their ethical and ideological convictions and their economic advantage as entrepreneurs—not in the beginning at least. The Lowell system united advanced technology, factory discipline, and conservative republicanism; and when it was eventually challenged, protest came on both economic and ideological grounds.

F. C. Lowell lived only until 1817, long enough to see the success of his Waltham experiment but before practical plans for the city that would bear his name had begun. Yet despite his premature death, he remained, in Nathan Appleton's words, "the informing soul, which gave direction and form to the whole proceeding."[29] To carry on his work, Appleton and Jackson selected as agent Kirk Boott, a trained engineer with an autocratic personality who had perhaps acquired his rigorous standards of discipline and strong class-consciousness in his service in the British army under the Duke of Wellington. They purchased the Pawtucket canal on the Merrimack in what was then the town of Chelmsford, together with four hundred acres of farmland, in the fall of 1821. Boott quickly set about the planning and construction of the industrial town according to F. C. Lowell's general conception, opening the first factory complex, the Merrimack Manufacturing Company, for production in September 1823. The company's six factory buildings were grouped in a spacious quadrangle bordering the river and landscaped with flowers, trees, and shrubs. They were dominated by a central mill, crowned with a Georgian cupola. Made of brick, with flat, plain walls, and white granite lintels above each window space, the factories presented a neat, orderly, and efficient appearance, which symbolized the institution's goals and would be emulated by many of the penitentiaries, insane asylums, orphanages, and reformatories of the period.[30] Beyond the counting house at the entrance to the mill yard stretched the company dormitories. Their arrangement reflected a Federalist image of proper social structure. The factory population of Lowell was rigidly defined into four groups and their hierarchy immutably preserved in the town's architecture. As chief agent for the corporation, most of whose stockholders resided in Boston, Boott and the other company agents formed the unquestioned aristocracy of the community; a Georgian mansion with an imposing Ionic portico just below the original factory in Lowell powerfully symbolized Boott's authority. Beneath this class stood the overseers, who lived in simple yet substantial quarters at the ends of the rows of boardinghouses where the operatives resided, thus providing a secondary measure of surveillance. In the boardinghouses themselves lived the female workers, who outnumbered male employees roughly three to one. Originally these apartments were constructed in rows of double houses, at least thirty girls to a unit, with intervening strips of lawn. Later, in the 1830s, as companies expanded and proliferated, the houses were strung together, blocking both light and air. These quarters were intended to serve essentially as dormitories and offered few amenities beyond dining rooms and bedrooms, each of the latter shared by as many as six or eight girls, two to a bed. Boardinghouse keepers were responsible for both the efficient administration of the buildings and for enforcing company regulations as to the conduct of the workers. Similar tenements were provided for male mechanics and their families. At the bottom of this hierarchy were the Irish day laborers, who built the canals and mills and made possible the continuing expansion of Lowell. Significantly, no housing had been planned for this group, and they lived in hundreds of little shanties next to a small Catholic church in an area called "New Dublin" and the "Acre." This early corporate insensitivity to the needs of the immigrant presaged Lowell's response to the great mass of immigrants later on.[31]

The adjustment of workers to factory life marked a critical juncture in America's transition to a mature industrial society. Many of Lowell's operatives had known long hours and hard tasks before in farms or shops; but the regularity and discipline of factory work were altogether new. They no longer labored at their own speeds in completing of a task, but to the clock at the pace of the machine. The employer aimed to standardize irregular labor rhythms and to make time the measurement of work. Thus the cupolas which crowned Lowell mills were not simply ornamental; their bells insistently reminded workers that time was money. Operatives worked a six-day week, approximately twelve hours a day, and bells tolled them awake and to their jobs (lateness was severely punished), to and from meals, curfew, and bed. Other factory owners also demanded long hours, even while they simultaneously claimed that the factory system had in large measure repealed the primeval curse "In the sweat of thy face shalt thou eat bread." In the hands of their operatives, they believed, leisure meant mischief; idleness at best; at worst vicious amusements, drink, gambling, and riot. Hence the resistance to shorter working hours throughout the nineteenth century and into the twentieth: work was a form of social control.[32] Lowell's managers shared this perception and wove it into the entire social order. They established an elaborate structure of social deterrents and incentives, insisting at all times upon "respectability" and defining it to suit their needs. Here the heritage of the Puritan ethic served employers especially well. Many Lowell women had been raised in a strongly evangelical atmosphere which placed heavy emphasis upon personal discipline and restraint. Injunctions to industry and the redemption of time pervaded their home communities, and their reading of popular didactic literature, from Isaac Watts's "How doth the little busy Bee," and Poor Richard's *Way to Wealth*, to the writings of Hannah More, reinforced these teachings. Company officials appropriated these values and adapted them to the imperatives of industrial capitalism. The Lawrence Company regulations, for example, stipulated that all employees "must devote themselves assiduously to their duty during working hours" and "on all occasions, both in their words and in their actions, show they are penetrated by a laudable love of temperance and virtue, and animated by a sense of their moral and social obligations."[33]

A policy of strict social control, implicit in the residential architecture, enforced this code of factory discipline. The factory as a whole was governed by the superintendent, his office strategically placed between the boardinghouses and the mills at the entrance to the mill yard. From this point, as one spokesman enthusiastically reported, his "mind regulates all; his character inspires all; his plans, matured and decided by the directors of the company, who visit him every week, control all." Beneath his watchful eye in each room of the factory, an overseer stood responsible for the work, conduct, and proper management of the operatives therein. Should he choose to exercise it, an overseer possessed formidable power. The various mill towns of New England participated in a "black list" system. A worker who bridled at employers' demands was charged with an offense of character, such as "insubordination," "profanity," or "improper conduct." Issued a "dishonorable discharge," she would be unable to find similar work elsewhere.[34] Supervision was thus constant. If the lines of social division occasionally relaxed on special occasions, it was only because the hierarchical authority of the community, which formed the basis of factory discipline, remained so indisputable.

In addition to these powerful institutional controls, corporate authorities relied upon the factory girls to act as moral police over one another. The ideal, as described by an unofficial spokesman of the corporation, represented a tyranny of the majority that would have made Tocqueville shudder. Declared the Rev. Henry A. Miles of Lowell, "Among the virtuous and high-minded young women, who feel that they have the keeping of their characters and that any stain upon their associates brings reproach upon themselves, the power of opinion becomes an ever-present, and ever-active restraint. A girl, *suspected* of immoralities, or serious improprieties of conduct, at once loses caste." As Miles approvingly described the ostracism, the girl's fellow-boarders would threaten to leave the house unless the housekeeper dismissed the offender. They would shun her on the street, refuse to work with her, and point her out to their companions. "From their power of opinion, there is no appeal." Eventually the outcast would submit to her punishment and leave the community. Even if, as one suspects, Miles overestimated the moral severity of Lowell women, his description nevertheless represented the official standard of behavior. On no account did employers wish to encourage independence of character, for it threatened the stability of the entire factory system.[35]

During its first two decades of operation, Lowell's reputation as a model factory town, offering economic opportunity in a wholesome moral and intellectual atmosphere, proved notably successful in attracting labor. Eager and intelligent young women flocked to the city, mostly from farms in New Hampshire, Vermont, Massachusetts, and Maine. Though their pay was not great and declined relative to the general economy over the years, manufacturing initially offered the greatest income of any occupation open to women at the time; domestic service in particular suffered as a result. Women came for manifold reasons: for money to assist their families, to support a brother's education, or to earn a dowry, and in some cases to gain independence from family life. Often Lowell women offered more romantic explanations as well: a failed family fortune, infidel parents, a cruel mistress, a lover's absence. As Lowell operatives reported their experiences and the community's reputation spread, many came for an informal education and the stimulation of their peers in an urban setting. In addition, company recruiters traveled through New England painting glowing pictures of the life and wages to be enjoyed at Lowell and collecting a commission for each young woman they persuaded.[36] With the construction of new factories and the rise of a middle class in the town to serve the needs of the enterprise, Lowell's population expanded rapidly: From roughly 200 in 1820, it climbed to 6477 in 1830, 21,000 in 1840, and over 33,000 in 1850.[37] For many young women away from home and family for the first time, the factory town appeared overwhelming at first, though most soon adapted to the new industrial environment and institutional life. Some even found the community rather snug and reassuring. With memories tinged by the nostalgia of old age, Harriet Robinson described the early days of Lowell as a life of "almost Arcadian simplicity," and Lucy Larcom recalled "a frank friendliness and sincerity in the social atmosphere," a purposefulness and zest for life which contrasted warmly with her early days as a child on the Massachusetts seacoast. Despite Lowell's swelling population and the lack of public parks until the mid-1840s, the town retained at least suggestions of a rural life. House plants in windows often gave corners of the mills the effect of a bower, and some of the overseers cultivated flower gardens behind the factories as well. According to Miss Larcom, "Nature came very close to the mill-gates . . . in those days. There was green grass all around them; violets and wild geraniums grew by the canals; and long stretches of open land between the corporation buildings and the street made the town seem countrylike."[38]

Gradually, most of these young women adjusted to the demands of factory life. Probably the greatest challenge confronting them was the machinery itself. "The buzzing and hissing and whizzing of pulleys and rollers and spindles and flyers"—as one ex-worker described them—often proved bewildering and oppressive for people completely unaccustomed to such devices. As they mastered their machines' intricacies, they learned to defy the noise and tedium by distancing themselves from their work through private thoughts and daydreams. Furthermore, before operatives were given more looms to attend and the machines speeded up in the mid-1840s, they often had long periods of idleness between catching broken threads. Regulations prohibited books in the mill, but women frequently cut out pages or clippings from the newspaper and evaded the edict. Others worked on compositions in their spare moments or spent the time lost in contemplation. Thus they attempted to give meaning to the time that their work denied and to cultivate a mental separation from their activities and surroundings.[39]

In the two or three hours they had remaining at the end of a long working day, and on Sundays, many Lowell women relentlessly pursued an education. They borrowed books from lending libraries, attended the lyceum at which Edward Everett, John Quincy Adams, and Ralph Waldo Emerson spoke, met in church groups, and organized a number of "Improvement Circles," two of which produced their own periodicals, the *Operatives' Magazine* (1841–42) and, most famous, the *Lowell Offering* (1840–45), and its successor, the *New England Offering* (1848–50).[40] Writers in these journals were self-conscious of their position as "factory girls" and eager to vindicate their reputations. As they endeavored "to remove unjust prejudice—to prove that the female operatives of Lowell were, as a class, intelligent and virtuous"—they offered impressive support for the Lowell system as a model republican community. Factory life at Lowell, a number of writers maintained, did not injure their health or degrade their morals. On the contrary, they asserted, the conscientious worker's "intellect is strengthened, her moral sense quickened, her manners refined, her whole character elevated and improved, by the privileges and discipline of her factory life."[41] To those who chafed against this regimen and thought of returning to the country, various authors replied that Lowell presented the most stimulating moral and intellectual climate, the most authentic republican community, in the land. Declared one woman in the *Lowell Offering*: "I believe there is no place where there are so many advantages within the reach of the laboring class of people, as exist here; where there is so much equality, so few aristocratic distinctions, and such good fellowship, as may be found in this community." A contributor to the *Operatives' Magazine* agreed: "We are, in fact, a truly republican community, or rather we have among us the only aristocracy which an intelligent people should sanction—an aristocracy of worth." While the stress of these remarks was more egalitarian than the conception of Lowell's founders, they effectively supported the existing system. The icon of the *Lowell Offering*'s title page depicted the symbolic landscape in which the operative stood: "the school girl, near her cottage home, with a bee-hive, as emblematical of industry and intelligence, and, in the background, the Yankee school-house, church and factory." With school and church, the factory thus formed a triad of republican instruction and uplift.[42]

As Lowell's fame spread in the 1830s, '40s, and '50s, countless visitors made the pilgrimage to the town, were conducted through its factories by representatives of the corporations, and emerged awe-stricken by its technological splendor and moral sublimity. Their rhapsodic testimonies overwhelmingly endorsed the policies of F. C. Lowell, his associates, and successors. Not only did the town appear to sustain the nation's highest standards of health, intellect, prosperity, and character; its success was such that in many respects it presented a model for American communities. In contrast to the myriad utopian experiments which spread through the country in the decades before the Civil War, only a few of which seriously attempted large-scale manufacturing, here was an experiment of the most practical sort, based not upon a notion of agrarian equality, but rather a paternalist technological order. Henry Colman, who was later so horrified with industrial conditions in Manchester, exclaimed after a tour of Lowell in 1836, "The moral spectacle here presented is in itself beautiful and sublime." In his view the elaborate machinery of these cotton mills, in which "each part retain[ed] its place, perform[ed] its duty," and worked in harmony, presented a model for human society. Colman implicitly admonished his fellow citizens, particularly the lower orders, to heed the lesson of the machine. All the wonderful results of Lowell were defeated "when even the most minute and the humblest part[s] of the machinery fail to perform their proper office," or the operatives relaxed their physical, intellectual, and moral energies.[43] In 1816 Walton Felch had endeavored to make a similar point in a poem significantly titled after the new time-conscious labor system, *The Manufacturer's Pocket-Piece; or the Cotton-Mill Moralized*. The discipline of the factory, Felch suggested, might provide just the salutary influence to keep republican spirits from running to excess. He chose a cotton mill as an illustration and carefully traced the moral lessons of control taught by each of the various elements; then Felch solemnly gestured to the whole and instructed his countrymen:

> Remark the moral order reigning here,
> How every part observes its destined sphere;
> Or, if disorder enter the machine,
> A sweeping discord interrupts the scene!
> Learn hence, whatever line of life you trace,
> In pious awe your proper sphere to grace.[44]

In appropriating the Puritan ethic to the demands of the factory, industrial spokesmen thus revitalized the conservative social and political implications upon which these values were originally based. What Colman and Felch described was essentially an industrial version of the Puritan doctrine of the calling, by which each person pursued his appointed vocation in the place which God had ordained.[45] Factory discipline would provide social discipline as well. The message undoubtedly seemed especially timely as democratic voices were challenging established authority. Thus they modified the Jeffersonian dictum: not of "the green world" but in the factory would Americans learn their place.

Thus it was with a certain vindictive satisfaction that Lowell officials welcomed Andrew Jackson to their showcase of industry and Federalist–Whig model of republican community during his northeastern tour of 1833. Jackson's Nullification Proclamation the previous year had only slightly modified the opinion of Boston and Lowell's leaders that he symbolized all the political principles and social forces they opposed; and they determined to mount such a display as to force him to acknowledge the magnitude of their achievement. Jackson would surfeit on their success. Declared Amos Lawrence, "We will feed him on gold dust, if he will eat it!" The companies and town organized an elaborate reception for the President. A procession of over 2500 young women from the textile mills, wearing white muslin dresses with blue sashes, carrying parasols, and bearing banners with the inscription, "Protection to American Industry," highlighted the event. "This, if nothing else," Lawrence's son Amos A. Lawrence noted with satisfaction, "was a splendid sight for the old general." After the procession, Jackson was escorted to the Merrimack Company mills to view "some of the works put in operation by the girls in their gala attire." Jackson understandably came away from his visit as impressed by the workers' fresh and refined appearance as by the technical aspects of production.[46]

More to the manufacturers' political taste, however, were two visitors who represented the Whigs' double-edged attempt to undermine Jackson's appeal by simultaneously representing frontier virtue and learned refinement, Davy Crockett and Edward Everett.[47] Crockett, an erstwhile Jacksonian, was wooed away by the Whigs and sent upon his own tour of the Northeast a year after the President's, where he denounced Jackson at every stop. He visited Lowell under the escort of Abbott Lawrence and gave the enterprise a simple backwoodsman's blessing. Compared to that of their enslaved sisters abroad, the service of Lowell women struck him as nothing less than enviable: here he discovered "thousands, useful to others, and enjoying all the blessings of freedom, with the prospect before them of future comfort and respectability." Some, lacking firsthand knowledge, whispered that these factories were no better than prisons, but Crockett assured his countrymen that Lowell was in fact an unusually healthy and happy community, a direct result of its democratic condition. "Respectability," philosophized the Tennessean, "depends upon being neighbour-like: here everybody works, and therefore no one is degraded by it; on the contrary, those who don't work are not estimated."[48]

Speaking at Lowell's Fourth of July celebrations for 1830, Everett also emphasized the community's achievement as inextricably related to its democratic character. At the wonder of the creation of the town from open fields in less than a decade, he feigned incredulity: yankee ingenuity and business sense must have been combined with some enchantment to make it possible. Yet the true genie, Everett averred, was not far to seek: "It is the spirit of a free country which animates and gives energy to its labor . . . makes it inventive, sends it off in new directions, subdues to its command all the powers of nature, and enlists in its service an army of machines, that do all but think and talk." Everett thus placed special emphasis upon the Revolution as the seedtime of American technology and hailed the community of Lowell as a fulfillment of that struggle. For the success of that experiment apparently offered invincible proof of what Everett and his generation so fervently believed, that technology, republican virtue, and prosperity might collectively advance. The specter of industrial degradation in the Old World need not prove controlling in the New. Although it might appear that "the industrial system of Europe required for its administration an amount of suffering, depravity, and brutalism, which formed one of the great scandals of the age," the presence of Lowell refuted any such charge with respect to America. Everett confidently affirmed that, "for physical comfort, moral conduct, general intelligence, and all the qualities of social character which make up an enlightened New England community, Lowell might safely enter to a comparison with any town or city in the land."[49]

American enthusiasm over Lowell was eminently shared by European visitors.[50] The town quickly emerged as the celestial countertype to infernal Manchester. By the 1830s it had become an obligatory stop on foreign itineraries, as distinctively a republican innovation as the American penitentiary, as established a landmark as Niagara Falls. Despite Lowell's international reputation, each traveler retained a European conception of factory towns which left him unprepared for what he saw. The dramatic natural setting along the banks of the Merrimack, nestled in the hills, with views

The discipline of the factory: Middlesex Company Woolen Mill.

reputedly as far as the White Mountains, no less than the crisp, clean aspect of the town itself, gave Lowell an air of "rural freshness" which dazzled foreign guests. As a result, each took his first glimpse of Lowell in amazement, even an air of disbelief. Viewing the city from a hilltop one winter evening, the Swedish novelist Fredrika Bremer compared it to "a magic castle on the snow-covered earth." Upon closer inspection she exclaimed, "To think and to know that these lights were not *ignes fatui*, not merely pomp and show, but that they were actually symbols of a healthful and hopeful life."[51] Alexander Mackay found himself searching in vain for "the tall chimneys and the thick volumes of black smoke" that characterized English manufacturing towns. Lowell's appearance of newness overwhelmed Charles Dickens in the early 1840s, so that it seemed to him created only yesterday. And the perspicacious

French engineer Michel Chevalier, who had earlier experienced "the delusive splendor" of the great Manchester mills, approached Lowell warily. His sense of pleasure at the town, "new and fresh like an opera scene," warred with his fear of its eventual decline, causing him to ponder, "Will this become like Lancashire?" Only gradually, watching Lowell operatives passing neatly through the streets and learning of their wages, did he wholly credit the enormous gulf between Lowell and Manchester.[52]

All European visitors were, like Chevalier, eager to determine for themselves the character and condition of the operatives. In reading the conclusions of these visitors, one must remember that they witnessed "institutional displays" rather than representative situations. They were informed and conducted through the mills by corporate officials, and hence met operatives under conditions most apt to encourage a sense of institutional solidarity.[53] Still, they were nearly unanimous in affirming that the life of the workers fully realized the promise Lowell extended by its physical beauty, and they substantially agreed with Everett that for prosperity, education, and character, its population might compare with any of its size. So enthusiastic was the Rev. William Scoresby with the appearance and moral condition of Lowell that he scuttled back to the manufacturing town of Bradford, England, to spread its gospel. There, in a series of lectures, he instructed his parish how to emulate Lowell's example and sought to whet their appetites for self-improvement by reading huge chunks of the *Lowell Offering*.[54] Charles Dickens, who was also impressed by that magazine, found the degree of culture attained by Lowell women challenged the whole European notion of the "station" and ability of the working class.[55] Another devotee of the *Offering*, Harriet Martineau, agreed. She believed superior culture constituted the principal difference between Lowell and English factory workers. "Their minds are kept fresh, and strong, and free by knowledge and power of thought; and this is the reason why they are not worn and depressed under their labors." Twice the wages and half the work could not in themselves do so much. After her American travels, Miss Martineau found herself in substantial agreement with Lowell's directors and American manufacturers generally that factories supplied "safe and useful employment for much energy which would otherwise be wasted and misdirected." By providing work and scrupulously insisting upon honest, decent employees, these employers had, in her opinion, to a great extent purified American society. Indeed, she believed, "A steady employer has it in his power to do more for the morals of the society about him than the clergy themselves."[56] Only Chevalier hinted at the cost of this rigorous institutional discipline and moral scrutiny in "a somber hue, an air of listlessness, thrown over society"; and even he acknowledged that, compared to the horrors of the English system, it was a small price to pay. The European response to Lowell was fairly epitomized by J. S. Buckingham when he concluded:

> Lowell is certainly one of the most remarkable places under the sun ... for I do not believe there is to be found in any part of the globe a town of 20,000 inhabitants, in which there is so much of unoppressive industry, so much competency of means and contentment of condition, so much purity of morals and gentleness and harmlessness of manners, so little of suffering from excessive labour, intemperance, or ill-health, so small an amount of excitement from any cause, so much of order and happiness, so little amount of misery or crime, as in this manufacturing town of Lowell, at the present time.[57]

Lowell's planners and directors might thus have felt deservedly proud of their accomplishment. For in Lowell and its sister cities —Chicopee, Holyoke, Lawrence, Manchester, Saco, and the rest— they had apparently built a productive, cohesive, and harmonious community based upon the earlier ideological fusion of technology and republicanism. Lowell promised not to compromise the nation's agrarian commitment, but, rather to supplement it, to strengthen the country economically, socially, and morally. The factory town ostensibly reconciled the myth of the American garden with a new myth of the machine.[58] Safely removed from Boston yet connected by the railroad, Lowell represented in the public mind a region in the middle distance, between city and wilderness. In this setting among the hills and on the banks of the Merrimack River, the town at once partook of the purifying influences of nature, yet—unwilling totally to submit to its siren song and reel as debauchees of dew—retained the beneficial discipline of the factory. The flowers in factory windows, so often noted by visitors, provided a fitting token of the community's premise, that an oasis of harmony and joy was attainable only through the maintenance of rigid moral standards and the fulfillment of hard work. One may protest that this represented a vitiated pastoralism, hardly worthy of the name; but this mythic fusion reconfirmed America's self-image as a natural yet disciplined republic and a land of abundance and opportunity. Prosperity and republicanism, the directors might have congratulated one another, had—despite John Adams's anguished cry—indeed been reconciled in a temperate and industrious community: This was the stunning achievement of Lowell. But was it?

Alongside the proud affirmations of company officials, the hosannas of industrial spokesmen and technological enthusiasts, and the admiring testimonies of European visitors, the 1830s and 1840s saw an insurgent attack upon the basic assumptions of the Lowell factory system and its conception of republican community. This assault was launched by members of the working class and their spokesmen, who, with the emergence of the labor movement, protested their oppressive working conditions and the hierarchical conception of society which sustained them.[59] Probably their sentiments were not shared by the preponderance of Lowell workers, many of whom shunned political opinions of any sort. But if these dissidents were a minority, they were nonetheless significant. Their very existence contradicted Lowell's image as a uniquely happy and harmonious community, and their arguments brought a radically different perspective to the institutionalization of the Lowell ideology and to the course of American technological development. Instead of remaining content in their station and allowing the social machinery to run smoothly, these workers rejected the notion that they shared a community of interests with mill-owners and called for the secret class war which was being waged against them to be fought in the open. The contrast between American and English factory systems did not appear to them so impressively distinct, and they were hardly inclined to join Whig politicians like Edward Everett in proclaiming Lowell as the fulfillment of the American Revolution and a model of republicanism.

Quite the reverse; the more extreme among them charged that the manufacturing elite had betrayed everything the Revolution stood for and were following in the footsteps of the luxury-loving and tyrannical British. Under the guise of humanitarian concern for the republic, they contended, Lowell's supporters were busily erecting a repressive new aristocracy.

One of the earliest and most important manifestoes of these dissident laborers was Seth Luther's *Address to the Working Men of New England* of 1832. Born in Rhode Island in 1795, the son of a Revolutionary War pensioner, Luther had worked for a time as a carpenter, then, after extensive travels in the West, returned to New England to devote his full energies to the cause of labor and reform.[60] His *Address* marked him as the region's leading unionist and became an important rallying cry for the whole labor movement. During the summer of 1832 Luther delivered the speech from Boston to Portland, Maine, and when published, it went through three editions. In it he attempted to speak for all laboring men, but particularly for operatives of cotton mills, among whom he claimed to have lived and worked for years. Largely self-educated, Luther began his address emphasizing the importance of broadly diffused education in a republic. The present system of manufactures, he contended, made this an impossible goal. The cruelty of the English factory system, which imprisoned its workers in ignorance, misery, and vice, and made them incapable of self-government, haunted his imagination. On innumerable occasions American manufacturers had declared that they too abhorred this system, but Luther ignored their protestations; instead he seized upon an exclamation of A. H. Everett (brother of Edward), "Witness the SPLENDID EXAMPLE of England," to confirm his charge that industrialists were fast imitating the English example at home. Lowell, he shuddered, might all too perfectly fulfill its title as "the Manchester of America." Already disturbances occurred there almost nightly, and riots were not uncommon. The wretched Irish hovels of that community represented, in Luther's view, the vanguard of "hundreds of thousands of the miserable and degraded population of Europe" who were enticed to this country, thereby depressing native American wages and degrading the entire factory population.[61]

The whole defense of the American factory system, Luther charged, represented a vast deception. "We see the system of manufacturing lauded to the skies; senators, representatives, owners, and agents of cotton mills using all means to keep out of sight the evils growing up under it. Cotton mills where cruelties are practiced, excessive labor required, education neglected, and vice, as a matter of course, on the increase, are denominated 'the principalities of the destitute, the palaces of the poor.' "[62] In truth, for scant reward American operatives worked even longer hours than their English counterparts; education and moral improvement inevitably suffered. The result was to create a caste system in America dividing rich and poor.[63] None of these tendencies and conditions was any accident. According to Luther, they sprung from the same menacing vices American colonists had earlier seen at work abroad, and especially from "*Avarice*," which ever "destroyed the happiness of the MANY, that the FEW may roll and riot in splendid luxury." Thus acquisitive lust, once largely confined to the shores of the Old World, now pervaded American manufactures, particularly the cotton mills. Luther pictured the daughters of mill-owners "gracefully sitting at their harp or piano, in their splendid dwellings, while music floats from quivering strings through perfumed and adorned apartments, and dies with gentle cadence on the delicate ear of the rich"; and all the while "the nerves of the poor woman and child in the cotton mill are quivering with almost dying agony, from excessive labor, to support this splendor."[64]

So long as such gross inequities existed, Luther contended, the American Revolution remained unfinished. The postulate of the Declaration of Independence, that "all men are created equal," he construed as a radical injunction against all social distinctions, and he called upon American workers, the source of all wealth, to unite and demand a just return for their labor. Against the common law doctrine then prevalent in the courts that combinations of workers were conspiracies against the state, Luther replied bitterly, "The Declaration of Independence was the work of a combination, and was as hateful to the TRAITORS and TORIES of those days as combinations among working men are now to the *avaricious* MONOPOLIST and *purse proud* ARISTOCRAT."[65] On the subject of his revolutionary rhetoric, Luther remarked in a similar speech in 1834, "Does any one think I use strong language, terms too severe, and unwarranted by the nature of the case? This language was used in the days of '76, and I contend there is more danger by far, now, than in those days. Then the danger was *apparent* to all eyes, now it is a *secret poison* in the body politic."[66] Yet it would be a mistake to conclude that Luther planned the wholesale destruction of American manufactures. On the contrary, he affirmed their necessity—one of the few points upon which he could agree with the early planners and directors of Lowell. Only, as they had questioned with respect to England, so he now doubted that in America "it is necessary, or just, that manufactures be sustained by injustice, cruelty, ignorance, vice, and misery; which is now the fact to a startling degree."[67] Ultimately, Luther too affirmed the possibility of a truly republican community, based on technology. But to achieve his egalitarian vision, vague as it was, involved more than simply attaining Luther's immediate goals of higher wages and a ten-hour day. It meant a fundamentally different social order, beginning with a complete restructuring of America's economic and class system. For the moment, Luther proposed to achieve this reformation through education. He denied that he intended any attack upon private property; but since he coupled this disclaimer with a demand that labor, "the most useful class," receive "a reward commensurate with its usefulness," mill-owners were not likely to breathe any easier. Undoubtedly they felt vindicated— and relieved—when, after years of political agitation, Luther's career as a radical finally ended in a lunatic asylum in 1846.[68]

Yet fervently as manufacturers may have wished to believe, Luther's ideas were not the isolated thoughts of a demented mind. They gained widespread currency within labor circles in the 1830s and '40s and were chanted right up to the mills of Lowell. In his address to the first convention of the National Trades' Union of 1834, Charles Douglas, who shared with Luther the mantle of New England's leading labor spokesman, pointed with alarm to the serious condition of female manufacturing operatives in America.[69] Lowell signified to him not a social model but a horrible example of the possible future of the nation. There four thousand women dragged out "a life of slavery and wretchedness. It is enough to make one's heart ache to behold these degraded females, as they pass out of the factory—to mark their wan countenances—their woe-stricken appearance." Such establishments, warned Douglas, were instruments of "a deliberate plot of the enemies of freedom and equality" to establish a new aristocracy. By luring farmers' daughters to work in the mills, these factories participated in a calculated assault upon the independent spirit of American free-men. The new generation of workers that would arise from this enfeebled stock would be incapable of further resistance. Douglas ended his speech calling for state legislation to limit the working day, but the rebellious attitude he expressed involved far more than a simple reform of factory hours.[70]

The fullest statement of the dilemma facing American work-ers came from another quarter than the speeches of labor orga-nizers, the pen of the brilliant, mercurial reformer, Orestes Brown-son. Alarmed and stirred by the threat of Whig victory in the Presidential campaign of 1840, he threw onto the scales of public opinion two long essays on "The Laboring Classes" in the July and October issues of his *Boston Quarterly Review*. Throughout the ante-bellum period corporate spokesmen contended that workers and shareholders rightly possessed a community of interests and even of ambition. According to Brownson, however, nothing like such social unity presently existed. The great evil of modern so-ciety, he declared, was "the division of the community into two classes" of capitalist and laborer. The social gulf between Abbott Lawrence's wife and daughters and the factory girls in his mills stood as great as that between nobleman and tenant, and their rela-tions less familiar. Despite individual exceptions, Brownson asserted, class lines were becoming increasingly rigid in America. No one ever grew rich as a factory operative; rather he endured a system of labor more oppressive and morally corrupting than even slavery. In words hardly calculated to make company officials sleep easier at night, Brownson declared, "The only enemy of the laborer is your employer, whether appearing in the shape of the master mechanic, or in the owner of a factory."[71] Manufacturing villages such as Lowell presented their fairest faces to the public eye; but he asserted a dark side existed nonetheless. The great majority of workers sacrificed health, morals, and spirit to the company with-

Inside the mill.

out bettering themselves at all. After a few years, they returned to their native towns with damaged reputations and possibly fatal illness, while their employers reveled in luxury and shed crocodile tears for the laborer.

To emancipate the worker from wage slavery, Brownson proposed a wholesale assault not only upon the factory system, but all the social institutions that stood behind it as well. He demanded the dismantling of the "priesthood" of conservative religion; a radical restriction of governmental powers; and abolition of the banking system, all corporations, all monopolies, all privilege, all inheritance of property. Only then might a truly Christian community based on technology emerge. Brownson did not expect factory owners to applaud his proposals. Indeed he envisioned a global, apocalyptic struggle between aristocracy and democracy. With mixed terror and anticipation, he heralded its coming: "It will be a terrible war! Already does it lower on the horizon, and, though the storm may be long in gathering, it will roll in massy folds over the whole heavens, and break in fury upon the earth.

Stay it, ye who can. For ourselves, we merely stand on the watch-towers, and report what we see."[72]

Bristling as it was with religious and social heresy, Brownson's essay raised a storm of controversy. Whigs seized upon "The Laboring Classes" eagerly and publicized it as the latest product of this Democratic Robespierre, who waited in the wings for Van Buren to be re-elected. Stunned by his recklessness, Brownson's associates for the most part muttered in their letters and diaries with disbelief.[73] Brownson's charges that Lowell women returned from the mills broken in health and corrupted in morals received strenuous objections from that city. Lowell's former mayor Dr. Elisha Bartlett denounced Brownson as a slanderer of Lowell women. Marshaling a mass of statistics of marriages, church attendance, polls on health, and mortality rates, he defended the city's factory environment as more healthy and upright than the mass of New England towns.[74]

An anonymous writer vehemently rejected Bartlett's assertions and figures, however, and supported Brownson's attack upon Lowell's mills as oppressive and exploitative institutions. Corporations, he contended, regarded operatives as themselves mere machinery, and strove to work them up to the limits of the company's technology. Far from protecting womanhood, then, the corporations were spawning a new female caste, like that in Melville's "Tartarus of Maids," so shaped by factory life that they neither desired nor could attain their "proper female sphere" as honorable wives and mothers in society. This critic called for laborers to organize for better working and living conditions and the diminution of working hours. Like other labor spokesmen of the period, he did not oppose technology itself, merely the restrictive use to which it had been put. Whose labor, he challenged the mill-owners, was "labor-saving" machinery really saving? Not the workers'. "The power of the Merrimack, by means of the machinery it moves, saves the labor of a *few*, that are born to wealth and influence, and adds many long hours of toil, to the burdens of those, who are doomed by necessity to labor for their bread." This writer took seriously the frequently invoked notion that technology was heaven-sent. Since machinery depended upon natural forces created by God, it belonged to no one exclusively. After

compensating the inventor for his work in a machine's develop-
ment, "all mankind have an inalienable right to share the benefits."
Here was a conception of technological community to send
F. C. Lowell spinning in his grave.[75]

The attack against the Lowell factory system gained momen-
tum in the 1830s and '40s and spread within the mills themselves
as Lowell women began to demonstrate on behalf of reform. In
1834 they participated in their first "turn-out," a demonstration
and short-lived strike. Their numbers were estimated from "nearly
eight hundred" (*Lowell Journal*) to two thousand (*The Man*),
varying with the sympathies of newspaper reporters. The workers
issued a proclamation asking the support of all "who imbibe the
spirit of our patriotic ancestors," and ending with the verse:

> Let oppression shrug her shoulders,
>   And a haughty tyrant frown,
> And little upstart Ignorance
>   In mockery look down.
> Yet I value not the feeble threats
>   Of Tories in disguise,
> While the flag of Independence
>   O'er our noble nation flies.

The immediate occasion of the "turn-out" was the announcement
of a 15 per cent reduction in wages, but it represented as well a
protest against Lowell's paternalism as unrepublican. As one of the
demonstrators announced, "We do not estimate our liberty by
dollars and cents; consequently it was not the reduction of wages
alone which caused the excitement, but that haughty, overbearing
disposition, that purse-proud insolence, which was becoming more
and more apparent."[76] Two and a half years later, in October 1836,
Lowell women struck against an increase in the price of board in
company houses, amounting to a one-eighth cut in wages. Again
they fortified their resolution by reminding one another of the
Revolutionary struggle against tyranny: "As our fathers resisted
unto blood the lordly avarice of the British ministry," they declared, "so we, their daughters, never will wear the yoke which has
been prepared for us."[77]

Thus, despite the founders' best efforts and most stringent regulations, the radical, egalitarian strain of republicanism they had hoped to suppress broke out within the fortress of Lowell itself. Like other disident workers throughout the nineteenth century, Lowell operatives returned repeatedly to the American Revolution and particularly to the Declaration of Independence to fortify and articulate their protest against what they regarded as a repressive social and industrial system. They insisted that since they were "created with certain unalienable rights," their labor could not simply be reduced to a commodity of which they were denied the fruits; human rights, "life, liberty, and the pursuit of happiness," took precedence over property rights.[78] Lowell's leading investors were also acutely conscious of the Revolution. Though Amos Lawrence, for example, was not born until 1786, so steeped was he in stories of that event that he felt himself "an actor in the scenes described," and a simple incident like the sound of a gunshot in 1843 instantly transported him back to the battles of Lexington and Concord in 1775. But the moral he and other manufacturers drew from the Revolution was very different from the workers'. As his biographer Freeman Hunt wrote shortly after Lawrence's death, "In all [the Revolution's] phases it was of a conservative character, aiming to maintain what was, and not seeking the development of fanciful theories. Our ancestors had no projects for the colonization of Utopia. The revolutionists were all on the other side."[79] So the revolutionists must have seemed again in the 1830s and '40s. Lowell's directors and other manufacturers were not about to surrender to these dangerous new visionaries. Prior to 1860 in Massachusetts not a single strike ended in victory for the workers or checked the reduction of wages.[80]

The workers were handicapped not only in the lack of union organization; the very institutional character of the Lowell factory system placed immense obstacles in the way of labor resistance. As George Fredrickson and Christopher Lasch have observed in considering the problem of resistance against another total institution, plantation slavery, "all total institutions are set up in such a way as to preclude any form of politics based on consent."[81] In such a situation the conditions for organized and sustained resistance were meager. The authority of Lowell's staff was directed not just at the workers' productive performance but their private activities and feelings as well; and traditional moral values were appropriated to reinforce the purposes and perspective of the institution. Political agitation not only smacked of "insubordination," but was also considered "unladylike" in the dominant culture. In this respect, extremely valuable allies to company management in quelling dissent were the much publicized journals that Lowell women produced themselves. The *Lowell Offering*, the *Operatives' Magazine*, and the *New England Offering*, though all nominally independent, served in effect as house organs, expressing solidarity between workers and management, and they were covertly encouraged by Lowell's directors.[82] Wishing to elevate the reputation of the factory girl and to prove her virtue and intelligence, these periodicals rigidly excluded criticism of factory conditions or management policies and resolutely presented cheerful expressions to the public. Occasionally, an editor would sharply rebuke those who violated their decorous image. "Constant abuse of those from whom one is voluntarily receiving the means of subsistence," Harriet Farley lectured dissident workers, was "something more than bad taste." If an operative really wished to improve her condition, Miss Farley suggested, she should leave the mills altogether. A character in Lucy Larcom's poem *An Idyl of Work* supported this point of view when she asked:

> Why should we,
> Battling oppression, tyrants be ourselves,
> Forcing mere brief concession to our wish?
> Are not employers human as employed?
> Are not our interests common? If they grind
> And cheat as brethren should not, let us go
> Back to the music of the spinning-wheel,
> And clothe ourselves at hand-looms of our own,
> As did our grandmothers.

This theme that rather than protest, the dissatisfied worker should go elsewhere and seek a separate peace recurred in the pages of the *Offering*. If wages should finally drop too sharply, another young woman grandiloquently declared, "I fear not for the crust of black bread, the suppliant voice, and bended knee; for then the inducement to remain will be withdrawn. Our broad and beautiful country will long present her spreading prairies, verdant hills, and smiling vales, to all who would rather work than starve."[83] In the last analysis, this supposedly "voluntary" character of Lowell and the fact that employment was temporary by design, encouraged cooperation between workers and management and mitigated against the formation of a class consciousness. High job turnover rates alone would have inhibited the development of a sense of solidarity among workers and of united opposition to their employers.[84] In addition, as women who regarded their work in the mills as transient rather than a career, most Lowell workers were not disposed toward collective solutions to factory abuses.

Other obstacles also stood in the way of Lowell's protesting workers. The Panic of 1837 and subsequent depressions threw an estimated one-third of American laborers out of work and seriously damaged the union movement. With jobs scarce, workingmen's organizations came to regard the system of female labor as doubly pernicious; not only did it harm the women themselves, it brought women in competition with men, thereby either throwing the latter out of work or reducing their wages. In light of this situation, the National Trades' Union suggested in 1839 that the solution to the female labor problem might be to keep women at home where they belonged. Women operatives, clearly, could no longer depend upon male labor spokesmen always to uphold their position.[85]

In spite of these impediments, however, resistance to the Lowell factory system gradually increased. By December 1844 Lowell's dissident workers, led by the redoubtable Sarah Bagley, had achieved sufficient strength to form an organization of their own, the Lowell Female Labor Reform Association. Its ranks swelled quickly: within three months it numbered three hundred members and by the end of 1845 it claimed six hundred workers in Lowell alone, plus branches in all major New England textile centers. Now workers were able to establish connections outside Lowell to the labor movement and hence to some degree to subvert institutional pressures. Immediately, they formed their own journal, *Factory Tracts*, and soon formed an alliance with the *Voice of*

Industry, a new labor weekly newspaper, and brought it to Lowell. Denouncing the *Lowell Offering* as "a mouthpiece of the corporations," these dissident workers powerfully inveighed against the oppressive character of factory life. Like Luther and Douglas, they pointed with horror to the specter of a degenerate race, spawned in the mills to serve as slaves to a manufacturing aristocracy. And as in earlier appeals to labor, they attempted to rally and organize workers by applying the language and lessons of 1776 to their own times. "Is not," the *Voice of Industry* asked, "the same secret fawning, devouring monster, wilely [*sic*] drawing his fatal folds around us as a nation which has crushed the freedom, prosperity and existence of other republics whose sad fate, history long ago recorded . . . ?" In such conspiracies, the paper charged, industrious and virtuous labor was inevitably targeted as the first victim. Every year its burdens grew more grievous, and the *Voice of Industry* demanded for all workingmen their God-given right to " 'life, liberty, and the pursuit of happiness.' " The fact that conditions of European operatives might be even worse, the paper argued, was essentially irrelevant: "The American workingmen and women, will not long suffer this gradual system of *republican* encroachment, which is fast reducing them to dependence, vassalage and slavery; because the English, Irish or French operatives are greater slaves, their condition more deplorable or English capitalists and task masters have the power to be more tyrannical and oppressive." An article in *Factory Tracts* similarly appealed to America's true nobility, its workers, to cast off the yoke of tyranny from about their necks before their country became "one great hospital, filled with worn out operatives and colored slaves!" It closed defiantly, "EQUAL RIGHTS, or death to the corporations."[86]

Such rhetoric reasserted the radical egalitarianism of the republican message which conservatives had been struggling to contain ever since the Revolution. References to the possibility of violent revolution formed a recurrent theme in the writings of Luther, Douglas, and other labor spokesmen of the period; now such phrases were being shouted outside the very mills of Lowell. How seriously is one to take them? Certainly no laborers were stockpiling arms and actively preparing for an insurrection. But neither should one dismiss such talk as a kind of verbal spice utterly without significance. In part, of course, it was intended to goad manufacturers toward reforms. Yet one cannot help but feel that it had an opposite effect: that such language only reconfirmed in Lowell's managers and stockholders their sense of the violence and chaos that would erupt should their institutional controls be removed. Of equal if not greater significance, this rhetorical violence also represented tacit admission of the immense gulf between labor reformers' vision of life in an ideal technological society, freed from the exigencies of industrial capitalism, and the formidable obstacles they encountered even to such minimal reforms as the ten-hour day. Indeed, the primacy of the ten-hour movement, as one historian has suggested, only reflects the extent to which workers accepted and fought within their employers' categories of time and work discipline.[87] From this point of view, one might speculate that the apocalyptic rhetoric of dissident laborers and other antebellum reformers signaled, not the weakness of American institutions, but their strength and the difficulty of gaining any real leverage for resistance. While continuing to affirm their faith in the ballot box, they summoned forth the image of revolution not only because it made another link in the carefully elaborated analogy between nineteenth-century workers and the American revolutionists, but because it provided a vague yet powerful metaphor by which the bleak conditions of the present might suddenly be wrenched to their conception of the future.[88]

Whatever efficacy the workers' protests had, then, was as symbolic rather than instrumental action; whatever gains they achieved were expressive rather than substantial.[89] Efforts at specific reforms encountered powerful resistance. Workers from several mill towns had petitioned the Massachusetts legislature for establishment of a ten-hour day and other factory reforms as early as 1842, with no response. The petition of 1600 workers from Lowell and elsewhere the next year met a similar fate. In 1844 a third petition was tabled until the next session; so that at last in 1845 a legislative committee held hearings on labor conditions for the first time. In the absence of an existing labor committee, the task was assigned to William Schouler, publisher of the *Lowell*

Courier and a staunch supporter of the corporations. Sarah Bagley and the Lowell Female Labor Reform Committee feverishly circulated new petitions to support the ten-hour cause and gained over two thousand signatures, half of them from Lowell. Two conflicting groups of witnesses then paraded before the committee. Spokesmen for the employers defended the healthful environment of the mills, while Miss Bagley and a group of operatives personally testified that Lowell workers endured overlong work days for insufficient pay to the detriment of health, mind, and spirit. To examine conditions firsthand, a portion of the committee went to Lowell, and they returned substantially impressed. Of their visit to the Massachusetts and Boott Mills, they reported, "The rooms are large and well-lighted, the temperature, comfortable, and in most of the window sills were numerous shrubs and plants, such as geraniums, roses, and numerous varieties of the cactus. These were the pets of the factory girls, and they were to the Committee convincing evidence of the elevated moral tone and refined taste of the operatives."[90] Thus Lowell's technological version of pastoral proved remarkably resilient; even in the midst of protest, legislators found confirmation of the essential rightness of the enterprise in a single whiff of a potted flower.

The committee as a whole affirmed the healthfulness of existing conditions and further shied away from a ten-hour law as jeopardizing Massachusetts' industry with respect to its neighbors. Paying left-handed tribute to Lowell's petitioners, the committee declared, "Labor is intelligent enough to make its own bargains, and look out for its own interests without any interference from us." While it acknowledged that hours should be lessened, mealtimes extended, ventilation improved, and other reforms instituted, the committee contended, "the remedy is not with us. We look for it in the progressive improvement in art and science, in a higher appreciation of man's destiny, in a less love for money, and a more ardent love for social happiness and intellectual superiority." The *Lowell Offering* could not have put it better. Protesting workers did gain official support in Dr. Josiah Curtis's report on public hygiene for 1849 and the minority report of the legislature's special committee on labor for 1850. Finally, the Lowell corporation voluntarily shortened the work day to eleven hours in 1853. But no ten-hour legislation was passed in Massachusetts until 1874.[91]

Lowell's leading investors, company agents, and other spokesmen did not reply to protesting workers directly; but beginning in the mid-1840s, they published several books, articles, and pamphlets defending Lowell corporations from their critics. One of the most important of these was *Lowell, As It Was, and As It Is* by Henry A. Miles, a Unitarian minister at Lowell. Plunging into the controversy over the condition of Lowell operatives, Miles presented a detailed justification of Lowell's corporate policies and their beneficent effect upon the community. One may gather the tenor of his argument and the generosity of his interpretation by noting what shortcomings in the Lowell system Miles conceded: that sickness *did* exist among Lowell's female operatives, for example, though not from overwork, but because they were sick to begin with, or neglected to wear a shawl or a good pair of shoes, or indulged in sweets, or sacrificed to save money, or caught cold in a stuffy church. In such a manner, Miles minimized or dismissed the various charges against Lowell corporations and concluded by lauding both their physical and "moral machinery." Lowell emerged in his account as a model republican community, a prototype destined to be copied throughout the region and the nation. "There have been laid for us here the foundations of a great success—a method of business well devised, and carefully adjusted part to part, a system of public instruction planned on a broad and generous scale, churches, Sunday schools, libraries, charities, numberless institutions to enlighten, guide, and bless this growing city." Little wonder that Amos Lawrence commended Miles's book to a friend as "well written and true."[92]

The Lowell corporations' position was quickly reinforced by pamphlets from two company officials, Thomas G. Cary, treasurer of the Hamilton Manufacturing Company and son-in-law of merchant, philanthropist, and Lowell stockholder Thomas Handasyd Perkins, and John Aiken, agent of the Lawrence Manufacturing Company at Lowell. They defended both Lowell's high moral achievement—singling out the moral supervision of company boardinghouses for special praise—and the "moderate" company profits. They took pains to refute charges against Lowell's leading investors that they had established themselves as a tyrannical new manufacturing aristocracy, " 'lords of the loom' fit companions only for 'the lords of the lash!' " Lowell's magnates made their fortunes before they entered manufacturing, Cary asserted, not at operatives' expense. "Nearly all the principal shareholders," John Aiken added, " . . . were once poor boys: as poor as any now working in the mills. By industry, prudence, perseverance, and good economy, aided by a kind Providence, they rose step by step to competence, and then to affluence." By cultivating these republican virtues, present workers might do as much. For nowhere, Aiken concluded, "is labor so richly remunerated; nowhere are the profits of labor so equally and equitably divided between the capitalist and the laborer." Such equality of opportunity made aristocracy impossible. Thus Aiken sought to counter workers' unrest by enlisting them in the growing cult of the self-made man.[93]

In response to workers' charges as well as attacks upon the protective tariff, Nathan Appleton himself joined in celebrating the opportunities of American labor and in reiterating the founders' vision of Lowell. A privileged European aristocracy exploited its workers, he agreed, but America as an agrarian republic honored the labor of all. Employers and employees were both in fact "workers," sharing a community of interests, and wages were such that every laborer might become himself a capitalist at will.[94] Appleton liked to gesture to his own experience as a case in point— even while adding that he never thirsted after wealth. As he recalled his journey from New Ipswich, New Hampshire, to Boston as a "poor boy" of fifteen to begin his career as a merchant, his description echoed—one suspects deliberately—Benjamin Franklin's famous account of his youthful entrance into Philadelphia two generations earlier.[95] Like Franklin, Appleton used the tableau both as a measure of the distance he had traveled in the course of his career and an assurance that other talented, industrious youths could follow in his footsteps. By paying homage to the creed of social mobility in this way, Appleton helped to channel egalitarian urges within the existing social order and to deflect potential social resentments back on the individual. The promise of mobility, in the hands of the wealthy and privileged, served ultimately as an instrument of social stability and control over the working class.[96]

Even as Appleton and others defended and memorialized the Lowell system, beginning in the 1840s life in the mills grew increasingly less attractive to young New England women. Previously, in the first two decades after F. C. Lowell returned from Europe transfixed by English textile machinery, enormous strides in productivity had been achieved through rapid technological innovation. After the 1830s, however, the rate of substantial innovation in textile manufacturing slowed considerably. Lowell and other New England textile mills faced increased competition in the sales of their goods and declining prices. Profits and dividends, after hitting their peak in 1845, sagged substantially with only a weak upturn in 1851–53 until the beginning of the Civil War. As a result, companies sought to increase output in relation to labor costs.[97] To boost productivity, New England textile manufacturers constructed larger and larger mills, powered by steam instead of water. The average size of a new mill, about 6000 spindles in 1835, swelled to 18,000 in 1847, and to 50,000 by 1883. At the same time cotton consumption per spindle also increased 50 per cent between 1840 and 1860. The work force in Lowell also expanded in this period, from 8560 in 1837, to 9235 in 1845, to 14,661 in 1855; but the increase in operatives was not proportionate to the increase in mills' capacity. Instead, work assignments were enlarged to include more machines. To stimulate production further, overseers were placed on a premium system and encouraged by bonuses to drive their workers harder, and the comparatively slow pace of the 1820s gave way to a more intense factory discipline. Operatives thus perceived a "speed-up" in which, as output rose, piece-rates dropped. As a result of these changes, workers' average daily earnings increased slightly in the later 1840s, but they did not compensate for wages and conditions that had become increasingly exploitative. The failure of the ten-hour movement added to discontent. When profits declined and new wage-cuts were announced in 1848, the turnover of operatives rose sharply; at one of the most prosperous companies, the Merrimack, average workers' tenure dropped to nine months. In the past, the city's mystique had proved a powerful agent of employee recruitment. Now, as mechanization spread to other industries, such as boots and cloth-

An icon of republican virtue: Frontispiece of *Lowell Offering*.

---

ing, and other employment opportunities rivaled the textile mills, Lowell's companies appeared in danger of losing their command over their labor force.[98]

Nevertheless, when the Lowell factory system was suddenly transformed in the late 1840s and 1850s, it was from quite another source than capitulation to workers' demands. At just the moment when company control of its traditional labor pool was growing shaky, Ireland's terrible potato famine that had begun in 1845 and the subsequent eviction of Irish peasants by their landlords triggered a massive immigration to America, over one and a half million people before the Civil War. As a major ship and railroad terminus, Boston received tens of thousands of these Irish immigrants. Upon arriving, however, they encountered a constricted social and economic life with little receptivity to foreigners. Nativist prejudice combined with the newcomers' lack of training and capital to shut them out of all but unskilled occupations. In such a position, Irish immigrants offered textile factories at Lowell and neighboring mill towns a ready and abundant supply of labor, and the incentive to accommodate demands of native workers diminished.[99] Almost immediately, Irish began to take the places of departing New Englanders in the mills; once established, their presence further hastened the flow of native workers to other jobs and discouraged the entrance of other New England women into the industry. Only 7 per cent of the operatives in Lowell mills were Irish in 1845, but by the early 1850s their proportion was estimated as one half, and it grew still higher year by year. Later in the century, the labor force would be supplemented by French-Canadians and other immigrant groups. Instead of predominantly single women, the Irish came as families. As adult males were discriminated against, Irish women and, increasingly, children went to the mills, the last receiving lower wages than ever. High turnover rates thus persisted. But if the work force was not immobile, neither was the work force in Britain; it was certainly not the kind of circulatory labor force able to enter and leave the industrial economy at will with which the Lowell system had begun. Most lived not in company boardinghouses to be supervised by alien authority, but with family or friends in the town at large. With a different culture, training, aspirations, and status from earlier Lowell workers, the Irish obviously did not immerse themselves in improvement circles and literary magazines. And the Lowell mills, which had eagerly received credit for the talents and accomplishments of earlier operatives, now found themselves without their trophies.[100] In less than a decade Lowell lost its prized population of well-educated and temporary New England women and with it the factory system's very rationale. Suddenly the basis that Lowell's founders and most ardent defenders had insisted constituted the principal difference between this city and English manufacturing towns and upon which its welfare would stand or fall—its lack of an established proletariat—was totally overthrown.[101]

Naturally enough, it was a fact Lowell's citizens found difficult to acknowledge. In a handbook and history of Lowell and its businesses published in 1856 when the social transformation was well under way, Charles Cowley insisted that Lowell had no permanent factory population: "This fact, and this alone ... has saved us from those evils of vice and ignorance, demoralization and misery, which have been engendered by manufactures in some cities in Europe." After a few years in the mills, Cowley maintained, Lowell workers fulfilled the Jeffersonian dictum and returned to agricultural pursuits in the "virtuous rural homes" in which they were born and bred. However, Cowley suddenly abandoned this idyllic picture to warn of the unspeakable evils that would result, should the city ever suffer the "curse" of a permanent factory population. Then "Lowell would become a foul blot upon the face of the country," and its once fair workers would be replaced by a barbaric horde. "Degraded to the level of the Indian Pariahs, their independence would be that of serfs; their liberty, that of prisoners; their leisure, that of workhouse paupers; their education, that of plantation negroes; their health, that of invalids; their chastity, that of harem women; and their wages, like the wages of sin, would be Death."[102] The extraordinary fervor with which Cowley imagined Lowell's possible fate implicitly revealed his awareness of what he would not openly admit. The character of the Lowell community had changed irretrievably, and celebrations inevitably gave way to jeremiads.

The early history of Lowell, like that of other institutional innovations during this period, thus revealed complexities far beyond the shaping powers or expectations of its confident founders. The factory system they established, no matter how benign in intention, was still based upon a hierarchical and manipulative model in which workers were passive agents, tied to the demands of machine production and industrial capitalism as a whole. Ironically, Lowell's very success in attracting educated and independent New England women meant that at least an outspoken minority would refuse to accept the management's conception of "republicanism," inherent in its strict factory discipline, and insist upon a true egalitarian order. Reliance upon temporary workers, in any case, only postponed the question of what accommodations industry should make for a permanent labor force. But the arrival of the Irish triggered not a new concern upon the part of corporate officials and a re-examination of their conception of republican community, but, on the contrary, increased apathy. In this respect, Lowell anticipated or paralleled a number of other total institutions concerned with education or rehabilitation, which in the 1850s and 1860s responded to the Irish presence by dampening their avowed reform expectations and blaming inmates' poor condition, not on an unwholesome environment as before, but on inferior heredity. Just as penitentiary, almshouse, mental hospital, and reformatory gradually moved from a role of rehabilitation to custodianship, so did the factory also move from the notion of a circulatory to a long-term population.[103]

By the 1850s the possibility of an integrated and harmonious republican community seemed further off than ever. Even while Lowell's praises echoed, its founders' optimistic vision lay tarnished, and its most ardent defenders were forced on the defensive. The problems of urban and industrial growth and social disorder which Lowell was established to correct had spread to the community itself.

3

Technology and
Imaginative Freedom

R. W. EMERSON

THE INTEGRATION of technology and republican civilization raised issues not only on an economic and social level, but also
on a cultural and imaginative plane. Both participants in the American Revolution and their nineteenth-century descendants agreed
that republicanism meant more than political liberty alone; it
heralded the liberation of the human mind and spirit. The achievement of American independence thus gave a strong and immediate
impetus toward the creation of a new republican culture. One may
see this impulse in various fields: in the lexicography of Noah
Webster, whose famous speller of 1783 attempted to create a new
national language, as independent of England as American politics;
in the poetry of Joel Barlow, who thought his epic poem *The
Columbiad* might surpass the *Iliad*, since his work would embody
republican principles and morals; and in the paintings of John
Trumbull, whose service as an aide-de-camp to Washington prepared him to paint an extensive series of pictures of the battles and
leaders of the Revolution and ultimately four war murals of heroic
proportions for the rotunda of the new Capitol in Washington.[1]

The cultural achievements destined for the American republic,
no less than her political and military victories, became staples of
patriotic oratory throughout the nineteenth century, and particularly in the ante-bellum period. Declared a Charleston speaker in
1819, "Here shall Poetry ascend to the sublimest, highest heaven
of invention; and her sisters Painting and Sculpture equal, nay, surpass all that Greece, all that Rome e'er boasted. Here, will an admiring world exclaim, is the land of 'Freedom, Arts, and Arms.' "
Thirty-seven years later, in Wisconsin, another orator was still
gazing expectantly toward the future: "May we not, relying on
the past as a pledge for the future, predict that this Republic, has,
under God, an exalted destiny to fulfill? May not here be realized
the golden age of which poets have sung, when literature and science shall flourish upon the same soil, the foster-children of government?" And still later, in 1871, another figure, this time more

impatiently and critically, attempted to summon the republican culture which would fulfill the promise of America's political institutions. "I say," wrote Walt Whitman in *Democratic Vistas*, "that democracy can never prove itself beyond cavil, until it founds and luxuriantly grows its own forms of art, poems, schools, theology, displacing all that exists, or that has been produced anywhere in the past, under opposite influences." Thus the "quest for nationality" in American literature and in American culture generally was a quest for republicanism as well, an attempt to realize in the nation's imaginative life freedom and power commensurate with her political achievement.[2]

As machine technology began to play an increasingly important role in ante-bellum America, the question of its relation to the development of republican culture and imaginative freedom grew critical. We have already seen how Edward Everett and other celebrants of American technology proclaimed an alliance between technological improvement and social and cultural progress. At the same time, numerous Lowell workers and labor spokesmen feared that oppressive factory conditions might ultimately reduce American operatives to an ignorant and enfeebled caste, slaves to the new industrial system. Debate over the problem of technology and imaginative freedom raged not only between social classes, but throughout the culture. Could modern technology expand the possibilities for creative power and human liberty, free Americans from drudgery and deadening routine, and bring them into closer communication with one another and with nature? Or might technology instead blunt people's imaginations and ethical sensibilities, alienate them from their environment, and perhaps even serve as a new instrument of tyranny? To learn more about the particular character of this debate, study of a single person may, paradoxically, serve better than a broad survey. And while no one may be said to be "typical" of an age, certain figures do by their very untypical sensitivity to ideas and experience powerfully express the concerns of their culture. In the middle third of the nineteenth century, the figure who can lay greatest claim to this position of centrality, particularly in regard to the problem of technology and imaginative freedom, is Ralph Waldo Emerson. In this sense, then, he may serve as our "representative man."

Emerson is qualified in this regard because, of all Americans in the nineteenth century, he was most concerned with the possibilities of the imagination in a democracy. He devoted himself not so much to politics directly as to "the politics of vision."[3] Suspicious of narrow adherents to political ideology, he cared about its broadest application. For Emerson political democracy was incomplete unless it led to full human freedom in a state of illuminated consciousness and perception in which each individual's full ethical and creative powers emerged. Separated from the American Revolution by a generation, Emerson and his contemporaries nevertheless looked to it as a pivotal event and prefiguration of a larger revolution in something·of the same way that English Romantics regarded the French Revolution. Of the latter, Robert Southey reminisced, "a visionary world seemed to open," and "nothing was dreamt of but the regeneration of the human race."[4] In a similar spirit, Emerson exclaimed in his "American Scholar" address of 1837, "If there is any period one would desire to be born in, is it not the age of Revolution; when the old and the new stand side by side and admit of being compared; when the energies of all men are searched by fear and by hope; when the historic glories of the old can be compensated by the rich possibilities of the new era?" "The American Scholar" has been hailed by generations of commentators as the nation's intellectual Declaration of Independence; the phrase is apt only if we understand that Emerson sought not simply national independence from "the courtly muses of Europe," but a declaration of independence of the individual imagination. As he expanded his theme the next year in the "Divinity School Address," he declared: "Wherever a man comes, there comes revolution. The old is for slaves. When a man comes, all books are legible, all things transparent, all religions are forms. He is religious. Man is the wonderworker." Such doctrines challenged the basic religious and social beliefs of Whiggish New England and were castigated by the uncrowned "Pope" of conservative Unitarianism, Andrews Norton, as "the latest form of infidelity." Asserting the divinity of nature and the morality of the universe, Emerson celebrated the individual who stood in primary relations to the world and refused to take life secondhand. He made a creative appropriation of the Kantian distinction (which he had learned through Coleridge) between the Reason, the revelation of the absolute, and the Understanding, or empirical world of sense. Believing that Reason was "potentially perfect in every man," Emerson determined to fight for its attainment against "that wrinkled calculator" the Understanding, which was forever pointing at custom and interest to argue that the visions of Reason were false or at least impracticable.[5] Thus he boldly attacked tradition, institutions, public opinion—all the external authorities other Americans were concerned to establish—which impeded the development of the individual imagination. Republican virtue, Emerson believed, would best be achieved through radical individualism. As John Dewey has observed, "Against creed and system, convention and institution, Emerson stands for restoring to the common man that which in the name of religion, of philosophy, of art and of morality, has been embezzled from the common store and appropriated to sectarian and class use." For this stance, Dewey heralded Emerson as "the Philosopher of Democracy."[6]

It is a title that would have surprised many of Emerson's youthful contemporaries, perhaps even in some moments Emerson himself. By heritage, birth, and education a child of Federalist Boston, Emerson always retained a portion of the political and social prejudices of that milieu. On the eve of Independence Day in 1822, for example, as a nineteen-year-old student, he found himself "croaking" with fears of democratic instability and excess: "Will it not be dreadful to discover that this experiment made by America, to ascertain if men can govern themselves—does not succeed? that too much knowledge, & too much liberty makes them mad." Although within a decade Emerson's youthful Weltschmerz and conservatism would be transformed into a fervor of transcendentalist revolt, the reservations of the critic of the democratic masses were never wholly stilled. One could, as Perry Miller has remarked, cull from Emerson's journals "enough passages about the Democratic party to form a manual of Boston snobbery." For example, "The favorite work & emblem of the Jackson Party," Emerson declared on one occasion, "is a Hog." However, if Emerson was repulsed by Jacksonian vulgarity, he revolted from Whig timidity and piety just as he did from the "corpse-cold Unitarianism of Brattle Street and Harvard College." Whiggism he pronounced "a feast of shells, idolatrous of the forms of legislature; like a cat loving the house, not the inhabitant." For Emerson was a democrat of the individual spirit, not of society-at-large, and he judged both political parties by his own high standard of individual liberty and ethics. As a transcendental democrat rather than a political partisan, he declared his faith: "Democracy has its root in the Sacred truth that every man hath in him the divine Reason or that though few men since the creation of the world live according to the dictates of Reason, yet all men are created capable of so doing. That is the equality & the only equality of all men."[7]

As the defender of this transcendental democracy, Emerson challenged both established religion and politics. He resigned his pastorate of the Second Church (Unitarian) of Boston in 1832 to become, in effect, a lay minister to the nation, and as such attracted a considerable following, particularly among the educated and discontented of his own generation. He both presented his views in voluminous writings and delivered them personally across the country in roughly 1500 lectures over the next four decades.[8] Testimony to the seriousness with which Emerson's message was regarded lies not only in the enthusiasm of his admirers, but in the vehemence of his opponents. However, the most powerful enemies to "the Divine Reason," as Emerson well understood, were not Andrews Norton and his Unitarian brethren, nor the Boston society that shut its doors against him. They lay instead in doubt, fear, habit, conformity, tradition—in all the forces that impeded the active and illuminated consciousness. Just as Lowell's protesting workers feared a social and economic counterrevolution, so Emerson and his transcendentalist allies fought against a counterrevolution of the imagination. Perhaps no major Romantic writer, English or American, attempted so strenuously as he to resist succumbing to disillusionment or a loss of creative and imaginative power, though even Emerson retreated from the radical egoism of his thirties to "build altars to . . . beautiful Necessity."[9]

It was from this perspective that Emerson considered the problem of technology and American culture. Characteristically, he did not approach the subject through disciplined, sustained analysis. Distrusting completed systems, he preferred to explore the range of possibilities in a continual process of redefinition, qualification, and variation of his argument. "I write anecdotes of the intellect," he declared; "a sort of Farmer's Almanac of mental moods." Yet "the fragmentary curve" that he traced in his work was more than simply the product of whim. It reflected a habit of mind that was essentially dialectical. Through dialectic, the "science of sciences," Emerson attempted to discriminate between and to reconcile the dualities that contended about and within him.[10] The superficial contradictions to which he was so notoriously indifferent were not a source of weakness but ultimately one of strength. His dialectical strategy was uniquely suited to expose the tensions of a culture struggling with alternatives in an effort of self-definition; as a mode of analysis of American culture it has been used by a number of leading critics up to the present day.[11] In particular, his approach enabled him to assess freshly the impact of American technological development. Ultimately, Emerson offered as profound an examination of the possibilities of republicanism in a technological society as did the experiment of Lowell, Massachusetts. Whereas the Lowell factory system precipitated a dialectical interchange over an industrial republican community, Emerson in his journals and essays constructed his own internal debate and dialectic on the character of imaginative freedom and moral development in a technological society. Each exchange in its own way tested the culture's faith in the direct relationship between technological and social progress.

Perhaps the best place to begin an examination of Emerson's response to technology is with his journal entry describing what was evidently his first train ride, a trip from Manchester to Liverpool in August 1833 during his visit to England:

> We parted at 11 minutes after six, & came to the 21st milestone at 11 minutes after seven. Strange it was to meet the return cars; to see a load of timber, six or seven masts, dart by you like a trout. Every body shrinks back when the engine hisses by him like a squib. The fire that was dropped on the road under us all along by our engine looked as we rushed over it as a coal swung by the hand in circles not distinct but a continuous glare. Strange proof how men become accustomed to

oddest things: the laborers did not lift their umbrellas to look
as we flew by them on their return at the side of the track.[12]

This passage both suggests some of the imaginative possibilities and
hints at some of the problems which Emerson found in nineteenth-
century technology. While impressed with the train's speed, what
particularly intrigued Emerson in this account was the effect of the
railroad on the individual's powers of vision. Vision was for Emerson the primary agent and symbol of imaginative inspiration.[13] The
radical dislocation of perspective of the railroad passenger fascinated him and left him straining for images to describe his experience. He compared the effect of rushing but apparently effortless movement alternately to swimming underwater "like a trout"
and shooting through the air "like a squib." The railroad suddenly
transformed the familiar countryside for its passengers into a
different medium, bathed in a red glare by the engine. At the time
of his train ride, Emerson was just recovering from the first great
crisis of his career, signaled by the death of his first wife, Ellen, in
1831 and the resignation of his ministry the following year. On the
brink of his own transcendental revolt, he found the railroad ac-
corded well with his assertion of philosophical liberation. He
quickly returned to the railroad yard to inspect the engines and
learn more of their wonders from the American inventor then in
England, Jacob Perkins.[14]

Yet even in the midst of his own exhilaration, Emerson had
noticed that the railroad workers by the side of the track had
grown strangely impervious to the train's presence. Though he
left the nature of their relationship to the railroad unexplored, the
problem he broached in rudimentary fashion here would become
a major theme in his own and in other American writing on technology: the way in which the amateur's capacity for imaginative
experience through technology is dulled and even transformed by
mechanical expertise and routine. Thus a generation later, in "Old
Times on the Mississippi," Mark Twain recalled how in "learning
the river" as a Mississippi steamboat pilot, he lost the capacity to
appreciate the romance and beauty that had thrilled him as a
novice. Similarly, Charles Lindbergh described his first airplane

flight in 1922 in language Emerson might have used: "Trees become bushes; barns, toys; cows turn into rabbits as we climb. I lose all conscious connection with the past. I live only in the moment in this strange, unmortal space, crowded with beauty, pierced with danger."[15] However, like Mark Twain, Lindbergh soon felt separated from his ecstatic initial vision by his very experience and technical expertise. He discovered he had exchanged the "poet's eye" of the novice for the utilitarian but dull perceptions of the professional aviator, the idealism of imaginative experience for the empiricism of daily routine. Emerson's first impressions of the railroad thus adumbrated a continuing issue in American culture. To preserve the poet's eye in an age of technology would become for him a central concern.

At first, however, the strange, disjointed perspective of the railroad passenger appeared to Emerson to aid the vision of the poet and even to support his transcendentalist philosophy. As he wrote in his journal in June 1834:

> One has dim foresight of hitherto uncomputed mechanical advantages who rides on the rail-road and moreover a practical confirmation of the ideal philosophy that Matter is phenomenal whilst men & trees & barns whiz by you as fast as the leaves of a dictionary. As our teakettle hissed along through a field of mayflowers, we could judge of the sensations of a swallow who skims by trees & bushes with about the same speed. The very permanence of matter seems compromised & oaks, fields, hills, hitherto esteemed symbols of stability do absolutely dance by you.

The view from the railroad, with its blurred flow of a continuous panorama, achieved and confirmed Emerson's desire to see nature as a whole in a panoramic mode, in which all objects were interrelated and in flux. This, he believed, was the inspired vision of Reason.[16] His railroad experiences thus led Emerson to exult in mankind's developing technological powers as new instruments of physical and imaginative freedom. When speaking in this vein, his language could approach the celebratory tone of the popular rhetoric of technological progress. In Emerson's first major expression of his transcendentalist position, for example, his essay *Nature* (1836), he gestured prominently to the new technology to justify his confidence in mankind's expanding possibilities. Discussing nature as "commodity" or material for man's needs, to support his point he drew an illustration that clearly bore the impression of his own railroad experience. Through "the useful arts," Emerson argued, man "no longer waits for favoring gales, but by means of steam, he realizes the fable of Aeolus's bag, and carries the two and thirty winds in the boiler of his boat. To diminish friction, he paves the road with iron bars, and, mounting a coach with a ship-load of men, animals, and merchandise behind him, he darts through the country, from town to town, like an eagle or a swallow through the air. By the aggregate of these aids, how is the face of the world changed, from the era of Noah to that of Napoleon!" In a similar spirit of enthusiasm Emerson wrote in his journal a few years later, "Machinery & Transcendentalism agree well. Stage Coach & Rail Road are bursting the old legislation like green withes." Technology appeared to him to be working on a physical level of a revolution analogous to that of transcendentalism on an intellectual and spiritual plane. Both represented explosive new forces directed against outworn conventions of thought and behavior. In what seemed to Emerson and his circle the primal and apocalyptic struggle between the party of the Past and the party of the Future, of Memory and Hope, of Understanding and Reason, he rejoiced that in modern technology he had discovered a potent and progressive ally.[17]

However, Emerson never critically capitulated to the plaudits of technology. Even in *Nature* he qualified his praise by observing that "the use of commodity, regarded by itself, is mean and squalid. . . . a thing is good only so far as it serves."[18] The question of the imaginative and ethical ends to which technology was put gave him a critical perspective on the popular equation of technological and social progress. As the millennial strain of Emerson's transcendental radicalism which emerged about the time of his first trip to Europe in 1833 shifted to a more tempered meliorism after the early 1840s, this emphasis became increasingly dominant.[19] Even prior to this shift, however, following his dialectical strategy, Emerson was grappling with the possible threat modern technology posed to the imagination. As his lyceum itinerary expanded and he became an experienced train traveler, even the railroad, symbol and ally in his quest for transcendental liberation, lost some of its poetic appeal and began to appear merely another instrument of repression in a new guise. "The railroad makes a man a chattel," he complained in his journal in 1840, "transports him by the box & the ton, he waits on it. He feels that he pays a high price for his speed in this compromise of all his will. I think the man who walks looks down on us who ride." Supremely confident of his ability to retain his interior stability, Emerson had earlier delighted in the way the railroad altered his perceptions of nature. But in league with commerce and industrial capitalism rather than with transcendentalism, the railroad and modern technology generally threatened to alienate man from nature and hence from himself. As Emerson wrote in 1839, "This invasion of Nature by Trade with its Money, its Credit, its Steam, its Railroad, threatens to upset the balance of man, & establish a new Universal Monarchy more tyrannical than Babylon or Rome. Very faint & few are the poets or men of God. Those who remain are so antagonistic to this tyranny that they appear mad or morbid & are treated as such."[20] Nevertheless, Emerson was determined not to be forced into a defensive posture nor to relinquish his position as the defender of the imagination over the forces of external constraint. In his lecture "The Poet" (first delivered in 1841, extensively revised and published in 1844) he attempted to redress the balance and to assert the power of the poet to control technology.

"The Poet" represented one of Emerson's most ambitious attempts to master the expanding technological environment and to integrate it with the natural world by the force of the creative imagination. The poet, as Emerson used the term, was not a mere versifier but "the man without impediment," who united the powers of experience, inspiration, and expression. He was the man of imagination, armed to combat the rush of facts which encumbered and tyrannized all mere men of sense and routine. Against the fragmentary tendencies of the age, the poet opposed a vision of organic unity. "As it is dislocation and detachment from the life of God that makes things ugly," Emerson declared, "the poet, who re-attaches things to nature and the Whole—re-attaching even artificial things and violation of nature, to nature, by a deeper insight—disposes very easily of the most disagreeable facts."

Among "the most disagreeable facts" for much of Emerson's audience and for the writers of his generation was the new industrial landscape; but he endeavored to reassure them on this point. "Readers of poetry," conditioned by traditional aesthetic categories, "see the factory-village and the railway, and fancy that the poetry of the landscape is broken up by these . . . but the poet sees them fall within the great Order not less than the beehive or the spider's geometrical web. Nature adopts them very fast into her vital circles, and the gliding train of cars she loves like her own." It was, of course, an avowedly idealistic solution, in which skeptics might see beneath Emerson's rhapsodic assertion of mastery a particular eagerness to dispose "easily of the most disagreeable facts." (Melville marked this passage and scoffed, "So it would seem. In this sense Mr. E. is a great poet.") Fitted into the poet's great Order or not, critics would argue, unpleasant aspects of technology continued to exist and to proliferate. But Emerson contended that to a balanced mind and controlling vision such as the poet's, it mattered nothing if mechanical inventions increased into the millions. Nature still could not be diminished: "The chief value of the new fact is to enhance the great and constant fact of Life, which can dwarf any and every circumstance, and to which the belt of wampum and the commerce of America are alike."[21]

Toward the end of his essay, however, Emerson confessed, "I look in vain for the poet which I describe."[22] His triumphant assertion of the power of the individual imagination to counter all opposition, to integrate all contradiction within a harmonious vision, proved difficult to sustain. Indeed, in the very urgency of "The Poet," one may see Emerson struggling to surmount the forces of limitation about him. The vulnerability of this claim of revolutionary power is, of course, a major theme in Romantic literature. In Emerson's own case, the death of his son Waldo in 1842 combined with other doubts and discouragements about his radical message, including the growing problem of technology, to form a watershed in his thought. Though the division is by no means absolute, after the early 1840s Emerson emphasized less the transcendent power of the individual than the beneficent destiny of the race, viewed nature as a source less of individual inspiration than of collective amelioration. His approach to technology correspondingly broadened from its focus on the inspired individual to a more sober consideration of technology's impact on republican society and culture.

One may see the beginnings of this change in an address Emerson delivered in February 1844, "The Young American." In it Emerson attempted to extend his treatment of themes he had dealt with earlier in "The American Scholar" and "The Poet." In many ways "The Young American" represented his most affirmative vision of the cultural possibilities afforded by technology. However, the affirmation was more strained and the promised benefits more austere than in his earlier writings. The problems technology posed for the development of republican culture, Emerson was coming to realize, were more complex than they had first appeared.

Emerson began his address heralding the emergence of an independent American culture and hailing the decisive effect of American technology in this liberation. In the past Europe had dominated the native imagination, and Americans had in effect been sent "to a feudal school to learn democracy." However, the transportation revolution was rapidly integrating America's disparate elements into union. "Not only is distance annihilated," he declared, "but when, as now, the locomotive and the steamboat, like enormous shuttles, shoot every day across the thousand various threads of national descent and employment, and bind them fast in one web, an hourly assimilation goes forward and there is no danger that local peculiarities and hostilities should be preserved." Emerson thus ratified the faith of James Madison and other supporters of the Constitution that the national unity destined by nature would be consummated by technology. Indeed, he went even further to suggest that American technology was forging a new national culture in which technology itself would occupy a prominent position. The perennial argument of American writers struggling to create a national literature, from Charles Brockden Brown to Cooper to Hawthorne, had been that the American artist faced a "poverty of materials." However, Emerson, who had earlier asserted the poet's power to see beauty in the common life about him and in the useful arts, gestured to a new field for American culture in the technological environment. Instead of bemoan-

The transcendental railroad: The Liverpool & Manchester Railway at Rainhill Bridge.

ing what Leo Marx has called "the machine in the garden," the railroad's intrusion upon the poet's contemplation of nature, he celebrated it. Railroad-building, he asserted, had "introduced a multitude of picturesque traits into our pastoral scenery." Among the elements of the picturesque, Emerson listed "the tunneling of mountains, the bridging of streams, . . . the encounter at short distances along the track of gangs of laborers, . . . the character of the work itself which so violates and revolutionizes the primal and immemorial forms of nature; the villages of shanties at the edge of the beautiful lakes, . . . the blowing of rocks, explosions all day, with the occasional alarm of frightful accident." All these and other aspects of the transportation revolution, he maintained, "keep the senses and imagination active."

His harmonious vision, already severely tested, grew still more strained as he lingered to consider the plight of the Irish workers who built the railroad. Worked excessively and ill-paid, these men were further exploited by inflated company prices for food and clothing. They lacked money and knowledge to determine their own futures, and as a result were shunted from place to place as their jobs demanded. Emerson admitted these points, yet urged as compensatory factors the unlimited opportunities America offered Irishmen and the superior schooling the nation extended to their children. He submitted, besides, that the Irish possessed a vivacity and good humor that exceeded the spirits of native Americans and let them bear their burdens lightly. In any case, he concluded (like some early visitors to Lowell), the hard regimen imposed by the railroad had distinct advantages as a means of social control. In his own village of Concord, he adjudged dull shovels to be "safe vents for peccant humors; and this grim day's work of fifteen or sixteen hours, though deplored by all humanity of the neighborhood, is a better police than the sheriff and his deputies."[23] The tone of these pages, uncertain at the outset, appears by this point out of control, as Emerson himself may have recognized, for he deleted his passages on railroad-building and the Irish when the address was reprinted. In the very effort of assuming his stance as the controlling poet, Emerson's ecstatic vision of human possibilities seemed to fade and he embraced conclusions he would at other times have repelled. Under certain circumstances, he now discovered, the fetters of technology achieved beneficial results.

Emerson's greatest cause for optimism in the unifying powers of American technology, however, lay in the way technology promised to educe the potentialities of the landscape and thus fulfill the nation's destiny as a pastoral republic. "The land," he affirmed, "is the appointed remedy for whatever is false and fantastic in our culture." And though railroads and factories had assisted the growth of cities and drained the countryside of "the best part of its population," Emerson looked to the railroad to reverse this process soon. By facilitating travel and communications across vast distances, the railroad had brought Americans into much more intimate relations to the continent itself. As a result, he estimated that the invention of the locomotive had accelerated America's development of the countryside by fifty years. "Railroad iron," he concluded, "is a magician's rod, in its power to evoke the sleeping energies of land and water." Providentially, just as another invention, the steamship, had "narrowed the Atlantic to a strait," and increased contact with Europe, the railroad had reasserted the orientation of American culture in favor of the West. Emerson eagerly foresaw the time "when the whole land is a garden, and the people have grown up in the bowers of a paradise."²⁴ Thus if he did not subscribe to Everett's celestial railroad, he had his own pastoral express. Technology, he was assuring his audience, would not abrogate but rather ensure America's claim to be "Nature's nation." By transforming presently poor and uncultivated earth, the railroad would guarantee the country's agrarian basis and fulfill America's promise as Eden.

Having presented the railroad as carrying America straight to the pastoral ideal, Emerson reminded his audience that the nation's journey thither would not always be easy. Like his countrymen on the eve of the Mexican War, he believed that America was "the country of the Future," guided by a beneficent Destiny. However, his was not the spread-eagle Providence of manifest destiny but a slow ameliorative tendency in nature. Nor was this nature the organic protector of the transcendent individual of Emerson's earlier essays. Instead, as he described it in harshly mechanical metaphor, nature emerged as a ruthlessly efficient administrator of the world-machine:

> Nature is the noblest engineer, yet uses a grinding economy, working up all that is wasted to-day into to-morrow's creation;—not a superfluous grain of sand, for all the ostentation she makes of expense and public works. It is because Nature thus saves and uses, laboring for the general, that we poor particulars are so crushed and straitened, and find it so hard to live. She flung us out in her plenty, but we cannot shed a hair or a paring of a nail but instantly she snatches at the shred and appropriates it to the general stock.

In contrast to his earlier radical egoism, as Stephen Whicher has observed, Emerson suggests in this passage the consistent Calvinist's entreaty, "Are you willing to be damned for the glory of God?" Accordingly, he willingly prepared to prostrate himself and his age beneath the juggernaut of the railroad and the rule of trade: "We build railroads, we know not for what or for whom; but one thing is certain, that we who build will receive the very smallest share of benefit. Benefit will accrue, they are essential to the country, but that will be felt not until we are no longer country-men."[25]

Despite such grim optimism, Emerson did not surrender his role as critic to become simply another cheerleader of American technological development. On the contrary, though he continued to enthuse over mechanical inventions and their effects at odd moments in his essays and journals, he grew increasingly concerned about technology's dominance in society. Instead of serving as a useful convenience, machinery threatened to become an enervating necessity; instead of freeing man and firing his imagination, it appeared to dull his sensibilities and even to enslave him. Emerson never commented directly on the example of Lowell, Massachusetts, but there is evidence that he found its aggressive industrialism particularly ominous in this respect. When Lowell corporations bought some land and water privileges at the outlet from Lake Winnipesaukee in New Hampshire, he bristled at their acquisitiveness. In the summer of 1847 he wrote in his journal: "An American in this ardent climate gets up early some morning & buys a river." He hires twelve or fifteen hundred Irishmen, digs a canal, and redirects the water to his mills. Next, "to give him an appetite for his breakfast," he builds a house and plots out a town with streets and lots, school, tavern, and church. With the aid of an engineer and an agent, he buys control of his water supply "and comes home with great glee announcing that he is now owner of the great Lake Winnipiseosce, as reservoir for his Lowell mills at Midsummer." Other observers, such as Edward Everett, had used such extravagant, almost tall-tale descriptions to celebrate Americans' power to mold nature and create new towns through technology in accordance with their slightest whims. Emerson, however, saw in all this industry a more sinister conclusion. "They are an ardent race," he continued, "and are as fully possessed with that hatred of labor, which is the principle of progress in the human race, as any other people. They must & will have the enjoyment without the sweat. So they buy slaves where the women will permit it; where they will not, they make the wind, the tide, the waterfall, the steam, the cloud, the lightning, do the work, by every art & device their cunningest brain can achieve."[26] The thrust of Emerson's satiric sketch was to throw the whole progressive defense of technology into question. Did technological achievements such as Lowell really signify advances in civilization or only the spirit of exploitation and lust for gain in a new guise? Emerson did not pursue the point here, but for the moment he had found, at bottom, mechanical invention and slavery to spring from the same shirking impulse, the same desire for reward without effort.

The question of the kind of civilization a technological nation produced continued to occupy Emerson as he sailed in October 1847 for a ten-month lecture tour and visit to England. "The most resembling country to America which the world contains," Emerson had called England at the conclusion of his earlier visit in 1833, and he approached it now in part as a way of gaining perspective on his own country and condition. As the ancestor of the United States and the most advanced technological civilization in the world, England encompassed in her development both America's past and possible future. Emerson's experience in England marked a critical stage in his thinking on technology. Like C. E. Lester, Henry Colman, and other visitors, he was shocked by the beggary of Manchester and reported it in his letters home as an object lesson in the blessings of American republicanism, bidding his daughters "thank God that they were born in New England" and spared the misery he saw abroad. In March 1848, as revolution broke out in France, he attended a Chartist meeting in London and heard talk of open rebellion. At the same time mobs roamed the streets, looting the shops.[27] Nevertheless, it was not these incidents that most affected his thinking, and they played little part in the book he distilled from his observations and travels, *English Traits* (1856). Instead, Emerson was most profoundly impressed by a larger pattern in which these events were only symptoms, the way in which English civilization as a whole had been transformed by mechanistic and technological values. *English Traits* sounded a double theme: it was both a hymn to the historic greatness of British character and a dirge, mourning its decay into a materialistic and industrial culture. Though his argument was scattered rather than systematically stated, Emerson traced how the erstwhile strengths of the English—their passion for utility, reverence for fact, animal vigor, indomitable will—had become separated from their original, noble aims and changed to weaknesses. From crests of the human spirit in the Middle Ages and Elizabethan period, "when the nation was full of genius and piety," the English had turned to the construction of machines and a mechanistic empiricism divorced from higher goals. English piety degenerated into tasteful paganism, speculative philosophy to political economy, poetry to ornament. The result, according to Emerson, was "the suppression of the imagination, the priapism of the senses and the understanding." In their relentless devotion to materialistic utility, the English risked sacrificing their own humanity. While the London *Times* machinized the modern world for its readers, Oxford, "a Greek factory," machinized the ancient; and Emerson detected "a slight hint of the steam-whistle" even in their modern muse. Man himself, Emerson declared, had submitted to become a product in Britain's political economy: "On a bleak moor a mill is built, a banking-house is opened, and men come in as water in a sluice-way, and towns and cities rise. Man is made as a Birmingham button." As men had been reduced to mechanism, machines had been elevated to personages: "Steam is almost an Englishman. I do not know but they will send him to Parliament next, to make laws."[28] A radical transvaluation of values was affecting the entire character of English life.

Symbolic of the transformation Emerson saw sweeping England was the eager development of machines to replace human labor. He reported how after Richard Arkwright had improved the spinning jenny it required only one spinner to do the work of a hundred men. Other improvements quickly followed, until laborers organized and struck to protect their wages. A group of textile manufacturers near Manchester determined to overcome this resistance. As Emerson reconstructed their thinking, they reasoned that "iron and steel are very obedient," and wondered, "whether it were not possible to make a spinner that would not rebel, nor mutter, nor scowl, nor strike for wages, nor emigrate?" Under their direction, Richard Roberts set about "to create this peaceful fellow, instead of the quarrelsome fellow God had made." He succeeded in making a self-acting mule, a spinning machine which effectually dispensed with adult workers altogether, requiring only a child in attendance. This the mill-owners hailed with hypocritical praise as an invention " 'destined to restore order among the industrious classes.' "[29]

Emerson's language suggests that he regarded such inventions as tantamount to the construction of an automaton or golem designed to circumvent human impulses and perform its master's bidding alone. As in Mary Shelley's *Frankenstein* and Melville's "The Bell-Tower," Emerson feared that the product of this perverse faith in mechanism would ultimately turn against its creator. "A man must keep an eye on his servants," he warned, "if he would not have them rule him. Man is a shrewd inventor and is ever taking the hint of a new machine from his own structure, adapting some secret of his own anatomy in iron, wood and leather to some required function in the work of the world. But it is found that the machine unmans the user." In describing the prototypic invention as man himself, Emerson was undoubtedly correct, though anthropomorphic imitation is often the most primitive rather than sophisticated form of invention.[30] Certainly his observation was true to his sense that each mechanical improvement correspondingly diminished the range of human expression. As mechanization was extended and perfected, he saw men's skills degenerating. Throughout his career Emerson paid tribute to the life of the farmer, who coordinated his day to nature and stood "in primary relations with the work of the world." But if, as he contended, nature never hurried, machinery operated tirelessly at an exacting pace. Its unnatural and exhausting regimen profoundly altered its attendants. Thus the true revolution Emerson saw impending in England was not a revolution of the working class but of machines themselves: "Mines, forges, mills, breweries, railroads, steam-pump, steam-plough, drill of regiments, drill of police, rule of court and shop-rule have operated to give a mechanical regularity to all the habit and action of men. A terrible machine has possessed itself of the ground, the air, the men and women, and hardly even thought is free." In Emerson's view, a mechanical totalitarianism had seized British society. Machinery had spread from a number of isolated units to a technological system until the nation had internalized its values in a new kind of civilization. Pressed by the need to describe the situation in its full intensity, he expanded the image of the machine to include all these internal and external senses and used it even as a radical metaphor for England herself. As he wrote Thoreau from Manchester, "Everything centralizes, in this magnificent machine which England is. Manufacturer for the world, she is become, or is becoming, one complete tool or engine in herself."[31]

The unprecedented physical power England possessed was for Emerson ironically misleading. In her devotion to mechanism both in technology and philosophy, she had starved the powers of the spirit. Following what Emerson believed to be the cycle of history, England had progressed from savagery to a golden age, gradually to enter a period of inevitable decline. Both in the broad outline of his historical theory and in the way in which he located the machine as symbolic of a modern epoch of unbelief, Emerson's *English Traits* anticipated *Mont-Saint-Michel and Chartres* and *The Education of Henry Adams*. Like Adams in France, Emerson approached England's ancient castles and cathedrals with a deep sense of humility, sometimes saying before one, "This was built by another and a better race than any that now look on it."[32] Indeed, his account of his visit to Stonehenge with Thomas Carlyle toward the end of the book dimly foreshadowed Adams's pilgrimage to Chartres. For like Chartres, Stonehenge stood as a symbol of unity in the universe in an age of faith, a unity that Emerson aspired to recapture for his own time and country, but that Adams a half century later could only hopelessly mourn.

What then were the implications of Emerson's examination of England's technological civilization for America? From one point of view, his analysis might bolster his confidence in the emergence of an independent American culture, free from the sway of Europe. For America was still in her exuberant adolescence with a continent before her, while England had done her utmost with limited resources and begun her decline. "An old and exhausted island," Emerson concluded, she "must one day be contented, like other parents, to be strong only in her children." Moreover, as he had declared in "The Young American," Emerson

shared the dominant conviction that America's republican principles and pastoral condition would guide her technological development in a course radically different from England's. Nevertheless, the degree to which England, "the best of actual nations,"
had capitulated to technological values renewed with greater
urgency than ever before questions about the impact of technological development upon America's republican culture.[33] Would
American ideals be degraded in a mechanical materialism as England's had been? Would the republican imagination also submit
to the tyranny of machines?

Emerson's visit to England increased his sensitivity to disturbing changes in his own country. Retaining the fresh vision of the
traveler upon his return, he discovered ominous signs that technology was rapidly transforming American values too as the national fascination with technological and material expansion threatened to obscure nobler ideals. Railroad, steamship, and telegraph
fed America's taste for immediate and vulgar success instead of
lasting achievement. "Everything is sacrificed for speed,—solidity
and safety," he protested in an 1853 lecture. And like Hawthorne
in "The Celestial Railroad," he suggested that the course down
which the nation rushed might be hell-bent at the last: "They
would sail in a steamer built of Lucifer matches if it would go
faster." Even the farmer, whom Emerson had long celebrated as a
sanative force in American life, grew tainted; and he soberly recorded Ellery Channing's view that "the Railroad has proved too
strong for all our farmers & has corrupted them like a war, or the
incursion of another race;—has made them all amateurs, given the
young men an air their fathers never had; they look as if they
might be railroad agents any day."[34]

Mindful of England's exaltation of prosperity at the expense
of principle, Emerson was particularly alarmed by the Compromise
of 1850. By this act Congress had attempted to preserve the delicate sectional equilibrium between slave and free states which had
become upset as California and New Mexico prepared for statehood. Under the compromise, California would be admitted as a
free state, the boundary dispute between Texas and New Mexico
resolved, and New Mexico and Utah organized into territories
with power to legislate on slavery for themselves. As compensation

to the South, a new and more rigorous fugitive slave law would be enacted and strictly enforced in both North and South, and only the slave trade, rather than slavery itself, would be abolished in the District of Columbia. Among the most celebrated supporters of the compromise was Massachusetts' Daniel Webster, who in his Seventh of March speech spurned anti-slavery arguments as a menace to the cause of Union. Meanwhile, behind the scenes his position was backed by prominent businessmen and their lobbyists eager to protect their network of investments.[35]

To Emerson the compromise and particularly the part Massachusetts played in it signaled the appalling moral torpor to which his countrymen had sunk in their devotion to technological materialism. The fugitive slave law, which implicated Massachusetts' own citizens in the enforcement of slavery, in his eyes unmistakably revealed his state and country's sacrifice of spiritual values to material success, so that Emerson found himself wrestling with the problem of prosperity and republican virtue which proponents of technology had promised to solve. What *was* the country to stand for if she sold her historic honor and virtue with such levity? If the sense of divine retribution for such apostasy had faded by the mid-nineteenth century, so that Emerson's countrymen were, in effect, "sinners in the hands of a Benevolent God," he had for all his transcendentalism enough of his ancestors' blood to respond with a jeremiad worthy of the occasion.[36] In an address at Concord in 1851 he bitterly lashed out at this latest testimony that "Things are in the saddle / And ride mankind." The pious phrases on behalf of freedom and religion men had been taught to mouth, he declared, now stood exposed as "hollow American brag." The nation's sense of injustice was blunted, and all the conveniences technology provided could not compensate for its loss: "I cannot accept," Emerson declared icily, "the railroad and telegraph in exchange for reason and charity. It is not skill in iron locomotives that makes so fine civility, as the jealousy of liberty. I cannot think the most judicious tubing a compensation for metaphysical debility." For all her corruption, Emerson warned, America could not hope to escape an ultimate reckoning: "It is the law of the world, —as much immorality as there is, so much misery. The greatest prosperity will in vain resist the greatest calamity."[37] Slavery, he declared, was the price of America's sins. (In this connection, his journal entry four years earlier on the Lowell corporations linking technology and slavery assumes special significance.) Emerson's hint of cataclysm would be fulfilled a decade hence in the Civil War.

Thus aroused and sobered, Emerson's stance toward technological civilization grew more critical. Though in the flower of his transcendental revolt he had celebrated technology as a stimulus to creative vision, in his later career he emphasized more its tendency to debase the imagination. Nineteenth-century proponents of technology never tired of contrasting the godlike powers of modern man with the afflicted condition of the savage. The progress of technology, they insisted, was synonymous with the progress of civilization. As Emerson grew more skeptical of such claims, he turned this argument on its head. He viewed early man not as benighted without machines but as possessing a greater vision for their absence: "The first men saw heavens and earths, saw noble instruments of noble souls; we see railroads, banks, and mills. And we pity their poverty. There was as much creative force then as now, but it made globes instead of waterclosets. Each sees what he makes." What mattered it that modern man could realize the fables of the ancients, Emerson would argue, if he could no longer dream them? From a world-maker man had become a tool-maker. This passage with its devastating satiric thrust properly recalls Thoreau. His experiment in a life of simplicity at Walden Pond constituted Thoreau's own radical attempt to discover the potentialities of the republican imagination which his neighbors had lost. Symbolically, he moved into his rustic house on July 4, 1845. While Emerson did not make so dramatic a gesture, his intent was the same. Granting that he would not turn back the clock of history, he nevertheless confessed nostalgia to experience life without impediments: "Machinery is good, but mother-wit is better. Telegraph, steam, and balloon and newspaper are like spectacles on the nose of age, but we will give them all gladly to have back again our young eyes."[38]

The answer for Emerson, then, was not to renounce technology, but to subordinate it to the imaginative and moral life. Against his fears that America was becoming dominated by technological materialism, he took comfort in the melioristic tendency of nature, which in the rough way he had described in "The Young American" would in time reconcile all things. "Nature uniformly does one thing at a time," he noted in his journal in 1848; "if she will have a perfect hand, she makes head and feet pay for it. So now, as she is making railroad and telegraph ages, she starves the *spirituel*, to stuff the *matériel* and *industriel*."[39]

The theme of Emerson's last major work, *The Conduct of Life* (1860), distilled like *English Traits* from lectures prepared in the 1850s, was how to place one's shoulder to the wheel of Necessity and assist this process of social evolution. If, like a nodding auditor in a lecture-hall, one catches only snatches of Emerson's message, for example, his declaration that "Man was born to be rich," one might wonder (with some commentators) whether Emerson had not been suddenly transmogrified into Russell Conwell. For Emerson boldly asserted the benefit of power and defended its pursuit by the industrial giants of the age. Certainly, he agreed, they were partial men, "monomaniacs" who pressed their schemes on all their acquaintances. But how, he asked, were America's factories built or railroads constructed "except by the importunity of these orators who dragged all the prudent men in?"[40] (Emerson himself had invested in several railroad companies and lost money on his stocks in the 1850s.) Though these boosters were ultimately sacrificed in their own projects, the public stood the gainer.

Emerson's purpose, however, was not to provide a rationale for the excesses of industrial capitalism, but, while acknowledging its contribution, to place it in proper scale. Power, he insisted in *The Conduct of Life*, must be purified by culture, by which he meant not the trappings of genteel respectability but proportion, balance, a disinterested and holistic sense of the purpose of life and an ability to subordinate the transient to the essential, the material to the spiritual. For this reason, while Emerson affirmed the necessity of industrialists, he reserved his greater respect for the inventor. The difference between them in Emerson's view was precisely the difference of culture. The former class achieved their goals by single-mindedness and brute energy. In the process of invention, however, Emerson could perceive " a spiritual act," which the industrialist merely repeated a thousand times. True, in a larger perspective the inventor himself was also an imitator, drawing toy models from "the arch machine," man himself. Still, he revealed by his example how "every jet of chaos which threatens to exterminate us is convertible by intellect into wholesome force." Perceiving the law and unity of the universe, he participated in nature's ameliorative effort. His greatest achievement, as Emerson made clear in a later lecture on "The Progress of Culture" (1866), was not utilitarian but metaphysical: "He has accosted this immeasurable Nature, and got clear answers. He understood what he read. He found agreement with himself. It taught him anew the reach of the human mind, and that it was a citizen of the universe." Emerson thus still found the correspondence between the individual imagination and nature which he had triumphantly declared in his transcendental radicalism, but where in the 1836 *Nature* the unity and purpose of nature culminated within the active observer, here the opposite was true. As the successor to Emerson's earlier emancipating figures of the scholar and the poet, the inventor also tapped the springs of power, but a power externalized. His task was to learn what Emerson called "the last lesson of life . . . a voluntary obedience, a necessitated freedom. Man is made of the same atoms as the world is, he shares the same impressions, predispositions and destiny. When his mind is illuminated, when his heart is kind, he throws himself joyfully into the sublime order, and does, with knowledge, what the stones do by structure."[41]

Stephen Whicher has called *The Conduct of Life* "ethics for the superior man," "a gospel for gentlemen," and certainly both Emerson's tone and his audience grew increasingly patrician in his later career. Yet the shift must not be seen as a social and philosophical acquiescence to State Street, a benign approval of the *status quo*. If Emerson's message grew more generalized and the call to greatness less revolutionary, this was partially because Emerson saw his ideal of a morally balanced and unified imagination become more inaccessible in a technological age. The task of Emerson's scholar, therefore, became not simply to think for himself but to provide moral guidance and inspiration for a society immersed in its own devices. "The age has an engine, but no engineer," he wrote in his journal in 1853. It devolved upon the scholar to fill this void. As Emerson described the scholar's duties in another entry the following year, they were to distill the wisdom of the past and to steady the unruly present, "to keep the first Cause in mind, to consecrate all to an aim, to be the engineer of the wonderful engine which the Nineteenth Century in million workshops builds." That the engine *was* wonderful Emerson still affirmed, but he remained concerned lest American civilization as a whole mindlessly submit to mechanical dominion as England had apparently done. For all Emerson's praise of technological power, he never relented in his insistence that it must be subordinated to the imagination; otherwise, it was simply a drug on the spirit. Far from providing philosophical shelter for a rampant industrial civilization, Emerson strenuously attempted to call it back to its original purposes. "The recurrence to high sources is rare," he complained in "The Progress of Culture." "In our daily intercourse, we go with the crowd, lend ourselves to low fears and hopes, become the victims of our own arts and implements, and disuse our resort to the Divine oracle. It is only in the sleep of the soul that we help ourselves by so many ingenious crutches and machineries." Emerson aimed, above all, to awaken the soul, to restore it to a sense of its own possibilities. In this effort no mechanical device could provide a shortcut. Yet he did not for this reason abjure technology. He only demanded that one learn to govern it morally as well as physically. As he declared in the essay "Power" in *The Conduct of Life*, "All the elements whose aid man calls in will sometimes become his masters, especially those of most subtle force. Shall he then renounce steam, fire and electricity, or shall he learn to deal with them? The rule for this whole class of agencies is,—all *plus* is good; only put it in the right place."[42]

Emerson's effort to "place" technology was a major theme throughout his career and a critical concern for the age in which he stood as America's exemplary thinker. Ultimately he offered only qualified affirmation. While granting the benefits of technology, he insisted that the imaginative and moral life which was the ultimate justification of a republic could never be reached by mechanical means. On the contrary, the development of Emerson's thinking from the 1830s to the 1860s reflects the growing difficulties he saw in the effort to attain the greatest imaginative freedom in what had indisputably become a technological civilization. Though he never retreated from his fundamental assertion of the possibility and necessity of such freedom, the pathway to it appeared increasingly clogged with obstacles. As he moved beyond that exhilarating vision of his first train ride at the outset of his transcendental revolt, technology seemed far less to aid the individual spirit and more to support the forces of social convention and restraint, less to illuminate a sense of idealistic purpose and more to encourage a life of materialistic pursuits and moral cowardice. Modern technology could not provide a substitute for republican ideals, he concluded; rather the proliferation of technology demanded firmer dedication to those ethical goals in order to stay in balance. The effort to attain this balance and to harmonize the nation's devotion to both technology and republicanism would be a continuing challenge for Emerson's contemporaries and successors.

4

The Aesthetics
of Machinery

By thud of machinery and shrill steam-whistle undismay'd,
Bluff'd not a bit by drain-pipe, gasometers, artificial fertilizers,
Smiling and pleas'd with palpable intent to stay,
She's here, install'd amid the kitchen ware!

So WALT WHITMAN imagined the Muse of the ages taking up residence in technological America. For Emerson, Whitman, and their society generally, a particular concern in the development of a republican civilization was the role technology would play in aesthetic ideas and experience. The rapid development and expansion of machine technology in the nineteenth century dramatically transformed the environment in which people lived, altered their perceptions of society and of nature, and presented spectacles unlike anything their colonial forebears had ever glimpsed. In reflecting upon that transformation, the modern reader is apt to see it simply in terms of aesthetic despoliation. His mind flashes to images of thick belching smokestacks of nineteenth-century industry, to coal-choked air and polluted streams, clanging machinery and hooting locomotives. No wonder he prefers to take his vacation at colonial Williamsburg rather than industrial Lowell! To speak of the aesthetics of such machinery, then, may well arouse disbelief. But nineteenth-century Americans beheld technology from a different perspective and with different assumptions than our own. Even Emerson, for all his vigilance against impediments to the imaginative development of the individual, saw technology as offering new aesthetic potentialities as well as challenges. Just as "in nature, all is useful, all is beautiful," so, Emerson believed, in art beauty and use needed to be combined and the distinction between the fine and the useful arts abandoned. Inspired by worthy ends, American technology might be the art of the future.[1]

The desire to fuse beauty and use, to see machine technology not simply as prosaically utilitarian but a source of aesthetic satisfaction, was widely shared by Emerson's countrymen. Defense and praise of modern machinery in nineteenth-century America characteristically included appeal to aesthetic criteria. Even habitual and highly trained observers frequently expressed delight in the spectacle of productive technology. Thus a correspondent for *Scientific American* at the Great Exhibition of 1851 in London enthused before a display of British locomotives: "Oh how I like to look upon those mighty iron arms heaving up and down or moving backwards and forwards at every heave of the steam giant's breast." Supplied with power from a boiler-house outside the exhibition hall, "the slumbering leviathans start like giants refreshed with wine, and throw their irresistible arms from side to side with terrific grandeur." Nathan Appleton and Francis Cabot Lowell were similarly seized with a sense of wonder when they sat "by the hour watching the beautiful movement" of Lowell's new power loom in 1814. Undoubtedly, it was more difficult to sustain this intense aesthetic pitch as a factory operative. Nevertheless, a contributor to the *Lowell Offering* declared that she had "sometimes stood at one end of a row of green looms . . . and seen the lathes moving back and forth, the harnesses up and down, the white cloth winding over the rollers, through the long perspective; and I have thought it beautiful." A more poetic writer in *Scientific American* could even entwine a deserted factory with nostalgic beauty to challenge Goldsmith's "Deserted Village": "There is no sight which conveys a deeper sensation of 'sadness lone,' than that of a factory, once jocund with the sound of an hundred voices, and the gleesome hurling of throstle and loom, standing tall, deserted looking, and silent. The once busy wheel, which gave motion to thousands of spindles, and hundreds of shuttles, stands gloomy and motionless, like a worn-out war steed. The bell that once clanged cheerily at the evening hour, no more calls out hundreds of gladsome toilers, gushing home through the factory doors, to enjoy the evening's recreation and repose."[2] When people could weep over a textile mill as they did over the "Old Oaken Bucket" and other objects of nostalgia of the period, technology may be said to have won a firm place in nineteenth-century popular art.

The pleasures of viewing machinery were by no means ac-

Artistic machinery: The steam engine "Southern Belle" at the New York Crystal Palace.

cessible only to the initiated nor was the aesthetic merely "functional." Spectators derived satisfaction watching great machines at work totally apart from economic and technical concerns and often despite their ignorance of mechanics. After visiting a ferry-boat engine room in 1848, Walt Whitman exclaimed in the Brooklyn *Daily Eagle*: "It is an almost sublime sight that one beholds there; for indeed there are few more magnificent pieces of handi-work than a powerful steam-engine swiftly at work! ... We do not profess to understand the tricks—or rather the simplicities of machinery." Twenty-four years later in the pages of *Scribner's Monthly*, a writer airily passed over the theory behind the Bessemer process, in which iron was fired to a molten mass, then blasted with a stream of air to burn off its impurities and convert it to steel, to describe its magnificent "scenic effect": "Our eyes are blinded by the brightness, yet fascinated by the play of colors that mark the process of the purification. The prevailing hue is a rose-tint of exquisite loveliness, lost in the dazzling whiteness when we look steadily, but reappearing as often as the eye is rested by looking away for a moment." Such industrial spectacles often fascinated visitors despite themselves. Even such an ironic observer as Nathaniel Hawthorne, who professed not to be ordinarily "interested in manufacturing processes, being quite unable to understand them," found himself captivated by a tour through an iron foundry in England in 1856. He watched entranced as "lumps of iron, intensely white hot, and all but in a melting state, passed beneath various rollers and were converted into long bars, to emerge curling and waving . . . like great red ribbons, or like fiery serpents wriggling out of Tophet." Trip hammers, he discovered, were "very pleasant objects to look at, working so massively as they do, and yet so accurately, chewing up, as it were, the hot iron, and fashioning it into shape, with a sort of mighty and gigantic gentleness in their mode of action." Exulting in the scene, he concluded, "What great things man has contrived, and is continually performing! What a noble brute he is!"[3]

Such descriptions of technology abounded in nineteenth-century America, so much so that one must qualify an observation of Tocqueville's. After his visit to the United States in the early 1830s, he declared in uncompromising terms that "democratic peoples . . . cultivate those arts which help to make life comfortable rather than those which adorn it. They habitually put use before beauty, and they want beauty itself to be useful."[4] Yet the converse was true as well: Americans wanted the useful to be beautiful. The reasons for this preference lay deeper than the factors of prevailing moderate wealth, lack of great fortunes, and desire for comfort which Tocqueville cited. Americans' intense aesthetic response to technology and their desire to discover beauty in utility were firmly rooted in republican values.

One may see the beginnings of this aesthetic disposition in the Revolutionary period as American leaders discussed what stance toward the fine and useful arts was appropriate for the young re-

public. Determined to protect the virtuous character upon which
the future of the nation depended, they regarded the fine arts with
a critical eye. Painting, sculpture, tapestry, and their sister arts
were intimately associated in American minds with princely courts,
priestly cathedrals, and all the luxury and debauchery which they
had sworn to shun. Rousseau's argument that the fine arts had
grown hopelessly decadent in the service of despots and should be
abandoned in their existing forms reinforced their concern. When,
in addition, they contemplated the vast expense that Europe's
artistic wonders represented and the immense challenge America
faced in maintaining her independence and settling a continent, it
is not surprising that they felt inclined to postpone any massive
campaign on behalf of the fine arts in the United States. "All things
have their season," counseled Franklin, "and with young countries
as with young men, you must curb their fancy to strengthen their
judgment. ... To America, one school master is worth a dozen
poets, and the invention of a machine or the improvement of an
implement is of more importance than a masterpiece of Raphael."
Striking a posture that would be characteristic of American attitudes toward the fine arts, he added significantly, "Nothing is good
or beautiful but in the measure that it is useful." Though Franklin
acknowledged that painting, poetry, music, and the theater might
properly gratify a refined society, they were objectionable in a
nation that could not afford them.[5]

Yet the aesthetic impulses of a nation are not so simply de-
ferred. Intellectual suspicion of the fine arts in America did not
necessarily mean emotional insensitivity, as the ambivalent reaction
of John Adams illustrates. Intuitively sensuous, with a keen appreciation of the physical qualities of life, Adams responded passion-
ately to the fine arts during his diplomatic tenure in the capitals of
Europe from 1778 to 1788. However, despite his deep emotional
response to the arts, indeed partially because of its very intensity,
he distrusted their influence and wished to circumscribe their role
in America. In an age that looked to the intellect to discipline the
tumult of the passions, the fine arts threatened to overwhelm the
senses, encourage frivolity and decadence, and mislead the public.
Controlled by a corrupt leader, Adams feared, the arts could become ready tools in the subversion of liberty. "Every one of the

fine Arts from the earliest times has been inlisted in the service of Superstition and Despotism," he cautioned Jefferson in 1816. The next year he warned the painter John Trumbull in still stronger language that "Architecture, Sculpture, Painting, and Poetry have conspir'd against the Rights of Mankind." If the fine arts had any legitimate function in the United States, Adams believed, it was the limited didactic role of commemorating America's Revolutionary achievement and inculcating republican virtue. He advocated art, not for art's sake, but for the sake of the republic.[6]

That the development of the fine arts in America was shaped by such concerns for its social and ideological functions has been well noted by recent historians.[7] Debate over how artist and architect might best embody national ideals and bolster the republican order continued from the end of the Revolution throughout the ante-bellum period. In the search for the style and symbolic images that would best express American values, artistic practitioners and critics turned first to neoclassicism and later, particularly after 1840, to Gothicism and a broad range of revivalistic modes. Throughout this effort, proponents attempted to justify particular forms in terms of their fitness for social function and expression. While certain figures, such as the sculptor and critic Horatio Greenough and R. W. Emerson, have been occasionally isolated from their broader intellectual context and hailed by modern commentators as precursors of twentieth-century "functionalism," in fact the necessity to combine beauty and utility, to marry form and function, was an imperative recognized by many of their contemporaries as the fundamental basis for a republican aesthetic. This belief, however, did not inevitably produce aesthetic consensus—let alone an anticipation of the Bauhaus principles esteemed by many modern critics; rather it provided a broad and flexible standard used to justify styles as diverse as Greek, Gothic, and Italian picturesque, materials as different as wood and cast iron.[8]

What has been overlooked or misunderstood is the way in which this concern for a republican aesthetic affected nineteenth-century Americans' response not only to the fine arts, but to the useful arts as well, particularly to the new machine technology. Horatio Greenough's admiration of how in the course of a ma-

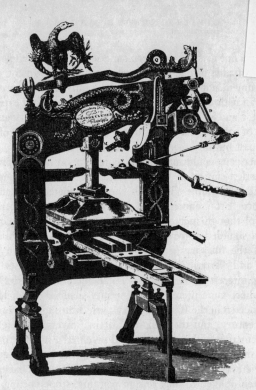

Republican iconography: The "Columbian" Printing Press.

chine's development and functional perfection its "straggling and cumbersome" shape gradually "becomes the compact, effective, and beautiful engine" has been cited again and again by scholars, as if he were unique in the nineteenth century in recognizing the aesthetic properties of machinery.[9] But Greenough did not need to teach his countrymen to appreciate the beauty of their technology, as their repeated exclamations of delight make clear. Nor was that his intention. Rather he wished to apply the aesthetic lesson of machinery to the higher beauty of the fine arts. For the public-at-large, however, machine technology itself helped to provide a solution to the aesthetic dilemma of a republican culture that demanded self-expression and symbolic statement even while it distrusted the fine arts. In a modern functioning engine, they beheld a product, capable of eliciting intense aesthetic enjoyment, that was the instrument not of decadence and tyranny but of a progressive, republican nation, the consequence not of idleness and expense, but of industry and ingenuity, the expression not of frivolity and weakness but of solidity and vigor, in short, the reflection not of Europe and the past but of America and the future.

Americans could more easily confer upon machinery a status and function similar to the fine arts, since their common origins were still discernible. Through the mid-nineteenth century the very word "arts" encompassed skilled crafts generally, including invention, though increasing use of the qualifying adjective *fine* or *useful* signified a growing sense of division. And while proponents of the fine arts gradually enforced a distinction between "artist" and "artisan," a number of American artists retained a strong interest in mechanics. These included Charles Willson Peale, whose vast interests and energies spanned the fields of art and science from the Revolutionary period through the early nineteenth century, and the sculptor Hiram Powers, who felt a lifelong fascination with machinery and invented a number of devices to assist his work. Less successful artists in the late eighteenth and early nineteenth centuries, still emerging from crafts backgrounds, were often obliged to turn to sign painting, stone cutting, engraving, and other artisan's tasks to earn a living. For their part, two of the most celebrated American inventors of the age, Robert Fulton and Samuel F. B. Morse, began their careers as painters. And Rufus Porter, the leading wall painter of this period, devised labor-saving devices for his art in the same ingenious spirit that he made over a hundred other inventions from washing machine to flying machine, scattering them about him like a "mechanical Johnny Appleseed" and publishing their plans in the various technical journals he founded, the most important being the *Scientific American*.[10]

Such connections helped substantiate the comparisons that observers of technology frequently drew between machines and the fine arts and their contention that the two sprang from related imaginations. Thus Edward Everett hailed modern machinery as resulting from "efforts of the mind kindred with those which have charmed or instructed the world with the richest strains of poetry, eloquence, and philosophy." A writer in *Scientific American* embellished this idea, arguing that "inventions are the poetry of physical science and inventors are the poets. Between the bards of machinery and the bards of literature, there is a strong resemblance; in fact, the same spirit of inspiration dwells in both—they only strike different lyres. . . . Who can tell of the dreamings—the wakeful nightly dreamings of inventors, their abstractions and enthusiastic reveries, to create some ballad or produce some epic in machinery." At the New York Crystal Palace Exhibition of 1853 visitors heralded the imminent union of science and art in technology and exuberantly applied artistic metaphors to machinery. In a walking-beam engine in the Machine Arcade, one observer discovered "the perfection of mathematical and artistic skill, . . . the most perfect product of Venus and Vulcan." The New York *Illustrated News*'s reporter came away from the exhibition exalted by the spectacle: "Science and art in an amicable wrestle for the smile of beauty; the loom and the anvil laughing out the jocund sound of profitable labor; the steam-engine snorting its song of speed; the telegraph flashing its words of living flame; the subdued ocean bridged with golden boats"—what did not all this promise for the future! To express the wondrous role technology was playing in modern history, he too drew upon the analogy of an epic and triumphantly concluded, "The Crystal Palace may be termed the Iliad of the Nineteenth Century, and its Homer was the American people."[11]

For all these public celebrations of the harmony between the fine arts and machinery, however, a sense of fierce competition remained. The machine aesthetic implicitly threatened the emergence of the fine arts in America, and some of its stronger advocates challenged the aesthetic priority of the fine arts directly. At a time when the artist was struggling to develop a sense of professional identity and social respect, to distinguish himself from the mere craftsman, so were the inventor and mechanical engineer.[12] The latter rebelled against the stereotypical view of the inventor as a simple, unlettered tinker and demanded for their own professions the kind of recognition that artists were gaining. Inventors and their spokesmen thus insisted that the distinction between technological pursuits and supposedly more lofty and refined artistic enterprises was at heart artificial, that mechanical achievements evinced creative intelligence as great in its way as did poetry or painting. Ultimately, however, they were not content with merely equal status but claimed the right to a superior position. For the inventor brought his countrymen, in addition to the pleasure of contemplating his imaginative achievement and the spectacle of mechanical performance, peace and prosperity as a result of his labors. According to his proponents, he stood both as a great creative artist and an immense practical benefactor. He thus offered, to an extent the fine artist could not match, both the beauty and utility requisite for republican art.

This argument was advanced by numerous technological advocates in the nineteenth century but none more boldly than Thomas Ewbank. Born of humble parents in Durham, England, Ewbank had begun his career as a sheet-metal apprentice at the age of thirteen. In 1819 he emigrated to America and quickly prospered as an inventor and manufacturer. Ewbank's success permitted him to retire from business at a relatively early age and to devote himself to study and writing on technical and scientific subjects. His efforts attracted national attention, and from 1849 to 1852 he served as United States Commissioner of Patents.[18] In both his official and other published writings, Ewbank took as his central theme the belief that God had made "the world a workshop," a gigantic factory, and had ordained as mankind's task to labor in it, converting raw materials into finished goods. Rejecting the idea that work was simply a punishment for the Fall, Ewbank enthusiastically summoned his fellow-men to fulfill the divine commandment as he interpreted it. To be an inventor and manufacturer was humanity's true destiny. As the truth of this gospel spread, technologists would at last receive proper recognition and serve as society's leaders and benefactors.

In articulating the universal plan, Ewbank reflected the inventor's horror of waste or inefficiency. With so much work to be done, non-utilitarian pursuits appeared to him to teeter on sin and blasphemy. "God employs no idlers," he declared, "creates none." "Like artificial motors, we are created for the work we can do— for the useful and productive ideas we can stamp upon matter."

The autodidact's suspicion and resentment of intellectuals and artists burned in Ewbank with particular intensity, and he disparagingly contrasted their attainments with those of the inventor. In the inventor stood one who refused to confine his wisdom to abstractions but employed it in tangible and productive creations. "In him the squeamishness of half-formed philosophers and of high-bred fashionables respecting manual and mechanical pursuits finds no sympathy, but terrible rebuke."[14]

What makes Ewbank especially interesting, however, is that he was not content merely to vent the self-made businessman's hostility to imaginative pursuits. In protesting the exalted stature of the fine arts, Ewbank did not mean to deny the importance of their function in society. He simply insisted that that function was better fulfilled by the technologist. Dethroning the romantic artist, he raised the figure of the romantic inventor in his stead. "Poems carved out of wood and forged out of metals" might rival the fine arts in profundity of inspiration and brilliance of design. Ultimately, because the inventor translated his conception into a material and useful medium, Ewbank argued, his work was more beautiful and sublime. As an "artist of the real," he towered over artists of the ideal, and Ewbank exulted, "A steamer is a mightier epic than the Illiad [sic],—and Whitney, Jacquard, and Blanchard, might laugh even Virgil, and Milton, and Tasso, to scorn."[15]

Ewbank aimed to divest the artist of his mantle not only as creator of the beautiful, but as a social leader and moral perceptor as well. Despite lingering suspicion of the fine arts' moral effect, opposition had thawed considerably by midcentury. Indeed, an increasing number of public spokesmen, including members of the once hostile Protestant clergy, were ordaining the artist as a kind of lay minister, to use his gifts to promote republican virtue and buttress the social order. Like John Adams, they accepted and justified art not for its intrinsic worth but for its social utility, as a didactic aid in the cause of republicanism.[16] It was this honored position that Ewbank attempted to wrest from the artist even at the moment of its bestowal. Deftly touching upon fears of art as leading to luxury and debauchery, Ewbank asked rhetorically, "How often are popular writers accused of pandering to the passions; but what contributor to the [useful] arts is a corruptor

Sightseers visit a Gothic engine room: U.S. Navy Yard Dry Dock, Brooklyn.

of morals? Like the works of the Divine Artificer, theirs tend to elevate, not to debase." The inventor, Ewbank contended, was in fact a practical moralist who, working in accordance with the principles of science, revealed the divine law, assisted in the spread of truth, and exposed error and evil. Proof of his moral efficacy lay in the history and present condition of the nations of the world. Declared Ewbank, "we invariably find those that excel in [the useful] arts most deeply imbued with moral principles—the foremost and most active in the benevolent enterprises of the age." Thus Ewbank concluded one of his reports to Congress as Commissioner of Patents urging its members to recognize inventors as the authentic voice of civilization and the vanguard of progress. With his keen jealousy of the fine arts, he was undoubtedly grati-

A Greek Revival steam engine, built by Harlan & Hollingsworth.

fied by Mississippi Senator Henry Foote's description of these reports as "more poetically grand, more brilliant, more fanciful, more Byronic than any of the most fanciful poems that Lord Byron ever produced."[17]

Though none was as audacious or as systematic in his arguments, other writers for *Scientific American*, a leading technological periodical of the day, echoed Ewbank's claims. "The fine arts," wrote one, "flourish best amidst a luxurious people, where wealth is concentrated in the hands of a few.... Objects of utility rather than objects of ornate ability, are the characteristics of American genius." Added another contributor, "Let them boast of ancient art who may—of ancient architecture, sculpture and painting— these are the signs of luxury and refinement and are not concomitants nor the signs of any nation's happiness or prosperity. A spinning jenny does more good to a country than a palace, and a steam engine confers more benefit than a temple."[18] Both writers insisted that they meant no disparagement of the fine arts, merely that in such a nation as the United States, these were clearly to be regarded as of secondary importance; but such assurances had a perfunctory tone. Among technological writers generally, the rarity of references to contemporary American writers and painters reveals their sense that art belonged to Europe and to another age. And unlike American artists of the period, who struggled desperately (and sometimes triumphantly) against this conviction, these reporters responded with hosannas instead of laments. Art's paucity in the New World signified to them the continuing health and vitality of the nation.[19] America, they congratulated one another, had not yet become a country of concentrated wealth, aristocratic pretensions, and moral decay. The united vocation and avocation, industry and art, of the republic would be technology.

From religious periodicals rose other voices to support technological writers in these beliefs. Resisting the revisionary attitude of some of his colleagues toward the fine arts, a contributor to the staunchly Congregational *New Englander* acknowledged that even technology might conceivably be used for evil purposes, but he contended that it was far less susceptible to abuse than the fine arts. For it was "unhappily true, that thus far on in the world's history, the loftiest genius, when acting upon mankind through the medium of painting and sculpture, and even music and poetry, has far too uniformly fostered the monstrous growth of corrupt and demoralizing principle, sensual and debasing passion." The difference lay in the fact that the fine arts, dealing with "merely *aesthetic* perfection," had no inner restraints. Technology, however, depended upon obedience to God's physical laws, which, like Ewbank, this writer saw as outward correlatives to the moral law. Challenging the trust that other clergy were confiding in the artist, he concluded that technology provided a firmer basis and truer support for Christian civilization.[20]

Perhaps the fullest tribute to machinery's practical and moral superiority to the fine arts was paid by John C. Kimball in the Unitarian *Christian Examiner* in 1869. In him distrust of art's effect and resentment of its lofty reputation were especially acute. Despite the fine arts' claim to the promotion of beauty and moral refinement, he argued that at bottom they were based upon sham. "It is not passion and power, beauty and sublimity, themselves, which they set before us, but their appearance. Their mission, or at least their means, is to deceive." Machinery, by contrast, aimed not at imitation, but at execution in accordance with natural laws. If faulty, it did not lure and dupe mankind; it simply refused to work. It therefore bore "something of the same relation to art that real life does to the stage, that the hero who performs a deed does to the actor who shows it forth." Thus contemptuous and suspicious of intellectual activity in general and the mimetic arts in particular, Kimball was not content with making the by now ritual comparisons of the steam-engine to the greatest epic, the spinning jenny to the finest sculpture. He professed to detect the stamp of inventors' superiority in their very appearance: "It is notorious that nearly all poets and philosophers, nearly all theologians, too, have something mean and little about them, all inventors something heroic and grand." That inventors were nonetheless denied proper recognition and accepted a career of suffering and sacrifice was for Kimball simply confirmation of their divine mission. As they endured their own metaphorical crucifixions, they carried on the work of Jesus, who had himself been a mechanic as His Father was an inventor. In invention lay religion. The progress of technology and Christianity were inseparably linked: to diminish class differences, to free men from drudgery, to sustain the high state of civilization requisite for progressive religion, Christianity depended upon technology. Machinery, then, was a gospel-worker: "The great driving wheel of all earthly machinery is far up in the heavens, has its force and direction supplied immediately from Omnipotence." Although they toiled in the workshop, inventors and mechanics thus emerged as the true artists and ministers of their age.[21]

The contention of these and similiar proponents of technology that machinery might supplant the function of the fine arts in a republic altogether was clearly a radical statement. But even if the general public did not push their conclusions to this extreme and grant machinery exclusive title as the art form of republicanism, they freely granted it an honored place. Recognition of the aesthetic and symbolic properties of machinery may be seen in the ways nineteenth-century Americans both designed machines and perceived them. Many of the same aesthetic values Americans demanded in the fine arts they sought also in technology. As a result, form followed not only function but also fashion and symbolic expression to a surprising degree in nineteenth-century American industrial design; so too, to an even much greater extent did observers' interpretations of mechanical form.

This point has been obscured by many twentieth-century commentators who, because they have themselves been profoundly unsympathetic to the dominant aesthetic of nineteenth-century Americans, have sought to identify instead an alternative machine aesthetic tradition that would provide a more congenial "usable past" on which to build. This effort has unfortunately shaped the interpretation of one of the most influential and in many ways most perceptive critics of nineteenth-century American technological design, John Kouwenhoven. In his book *Made in America* Kouwenhoven attempts to define a "vernacular tradition" in the American arts, a tradition he finds first and most clearly displayed in American technology. Citing numerous contemporary reports of the boldness, simplicity, and economy of American tools and machines, Kouwenhoven argues that American technology constituted a kind of folk art, the attempt of a democratic society to create aesthetically satisfying forms in an industrial environment. Even he must admit, however, that this vernacular tradition in technological design did not exist unalloyed. He gives evidence that a persistent, countervailing tendency to ornament American machinery developed and shaped the nation's bold engineering. Kouwenhoven identifies this "cultivated tradition" with Europe; but he acknowledges that it was shared by the majority of Americans, who often failed to see in the indigenous beauty of technological materials the substance of art.[22] Ultimately, Kouwenhoven offers little support for the contention that many Americans in the nineteenth century admired machines specifically for their functional beauty. Although they appreciated sound and efficient engineering, their aesthetic claims were based on broader grounds than these.

Indeed, much of Kouwenhoven's own evidence points to a conclusion opposite to the one he draws: as American mechanical engineers and manufacturers attempted to make their products "artistic," they frequently resorted to ornamentation. Those who decried this tendency, on the other hand, were not untutored representatives of a vernacular tradition, but the American and British engineering elite. One may see this struggle between philosophies of industrial design in the leading technological periodicals of the day, particularly in their comments upon the various world's fairs from 1851 to 1876.[23] While leading engineering critics usually demanded a utilitarian appearance in machine exhibits, many manufacturers preferred to display works with a bolder and more overt artistic statement, and to these especially fairgoers responded. A center of attention at the New York World's Fair of 1853, for example, supplying the power for the industrial exhibits, was the elaborately canopied steam engine, "Southern Belle," made by the Winter Iron Works, Montgomery, Alabama. Its pretensions as ornament caused one reviewer to complain, "The Southern Belle, running without any labor, is true to the name *belle*—very showy, and (at present) very useless. No shop would ever dream of making or buying such an engine for use. It would keep one man busy the whole time just to keep it bright and clean."[24]

Similarly, one of the hits among visitors to the Paris Exposition of 1867 was the locomotive "America" from the Grant Locomotive Works, Paterson, New Jersey. The first American locomotive to be displayed at a European exposition, the "America" was intended as a statement of both republican engineering and republican machine art. Mechanically, the engine was a typical American eight-wheeler, built lighter and more flexible than European engines, to accommodate the light but rough American roadbeds, with their steep grades and sharp curves. However, this sound example of American engineering was given dramatic artistic expression by its exquisite finish. The boiler, smokestack, valve boxes, and cylinders were encased in German silver (an alloy of copper, nickel, and zinc), and the cab made of ash, maple, black walnut, mahogany, and cherry. The elaborately scrolled tender made the patriotic appeal explicit: it bore the name "America," the Arms of the Republic, and a portrait of General Grant, hero of the Civil War and patronymic associate of the locomotive company's president, D. B. Grant. The New York *Tribune*'s Paris correspondent cheered the "America" as the "most majestic contribution" the United States had at the fair, and an international jury awarded it a gold medal, the highest prize given to any locomotive in the exposition. However, reports conflicted as to whether this last triumph came despite or because of the locomotive's decorative finish. *Scientific American*'s correspondent reported that European engineering critics were generally offended by the "America's" display, and he shared their reactions. American locomotives were already "showy enough without any additional embellishments." Yet, he observed, the "America" represented a trend in United States machine exhibits that all but the best makers followed. "Fancy painting seems to be thought a necessary qualification for an exhibited article, while all the foreign machinery is painted in perfectly plain colors."[25]

Artistic renditions of machinery by American manufacturers continued to flourish in the 1870s despite the objections of a number of engineering critics, both European and American. At the Vienna International Exhibition of 1873, a British official complained, "The besetting fault of the second-class American houses is the attempt at finery, both in regard to form, and painting and ornamentation."[26] At the Philadelphia Centennial Exposition of 1876 the same tendency persisted. A critic for *Scientific American* found himself appraising a semi-portable steam engine with bright blue boiler, maroon base, black pipe, and bronzed rivet heads, "a combination of which nothing could be more inappropriate or in worse taste in a piece of machinery." In desperation, the reviewer was tempted to codify a rule, that "a highly colored and fancifully ornamented piece of machinery is good in the inverse ratio of the degree of color and ornament." Similarly, the British periodical *Engineering* frankly confessed its puzzlement and irritation at the spectacle of American ornamented machines in Philadelphia:

> Ornament as a characteristic of engineering practice in America, is one of which it is hard to speak comparatively. It is a

President Grant and Dom Pedro starting the "sublime" Corliss Engine.

peculiarity almost wholly their own, and it is extremely difficult to understand how among a people so practical in most
things, there is maintained a tolerance of the grotesque orna-
ments and gaudy colours, which as a rule rather than an exception distinguish American machines. There is something so
anomalous in attempting to ornament a machine which is to
be covered with grease and dirt and operated where no one
cares for or ever notices decoration, that the practice must do
much to hide good qualities when they exist.[27]

Why then did Americans decorate machines? Why was the
machine aesthetic by which most manufacturers wished to be
represented at world's fairs not, as Kouwenhoven's argument
would suggest, one of functional simplicity, but one of embellish-
ment? One may suggest several reasons. First, there are the historic
origins of American machine design. In the early nineteenth century machine frames were commonly made of wood, and they
were designed by or in the spirit of cabinet makers. Stylistic trans-
ferences naturally resulted, and machines bore the decorative and
architectural details of furniture and household trim.[28] Still, this
circumstance alone, which was also the case in England, would
hardly explain the persistence of decorative motifs in metal for the
better part of the century despite continual criticism from the
leading engineering authorities of the day. Instead, one must consider machines' appearance in relation to both their immediate pur-
chasing audience and their broader social milieu. An important
factor in the aesthetic perception of nineteenth-century technology
was the increased complexity of machinery. As engineering grew
more intricate and machines adopted casings and housings that
masked their working parts, it became difficult for all but the most
knowledgeable observers to judge their performance readily.[29]
Thus, the inclination to view machinery as aesthetic spectacle was
accentuated by the inability of many viewers to evaluate machines
purely on a mechanical basis. The diverging responses of engineering critics and the general public before ornamented machine exhibits at the world's fairs reflected this growing gulf of expertise.
Evidently, some buyers of machines, hard pressed to appraise a
product mechanically, also weighed factors of appearance heavily
in making their purchases. "The majority of those for whom ma-

chines are constructed," argued the Englishman Samuel Clegg, "cannot enter into the merits of their internal action or of their comparative performance; and therefore, not being willing to yield the right of opinion, judge them from 'outside show,' as they would of a picture, or statue, where the only aim is to charm the eye, or excite pleasurable sensations in the mind, by the representation of Nature in her more sublime effects."[30]

Confronted with this situation, the manufacturer could choose between two basic strategies of machine design. He might follow Clegg's implicit suggestion and attempt "to charm the eye, or excite pleasurable sensations in the mind" in the decoration of his machinery. Or he could emulate the outstanding American machine-maker of the nineteenth century, William Sellers, who departed from accepted designs and aimed for appearance which focused attention and reinforced confidence in the product's engineering qualities. To this end Sellers eschewed ornament of all kinds, introduced "machine gray" paint in lieu of bright colors, and stressed simple, flowing surfaces which visually complemented the machine's function. It must be stressed that Sellers's effort involved a much more sophisticated concept of design than the idea that "if a machine was right, it would *look* right." Using simple geometric shapes, he gave his machines a formal unity that suggested their functional integrity, not an inevitable result if he had concentrated on mechanics alone. Furthermore, a machine would presumably work as efficiently painted bright colors as dull. Intuitively or not, however, Sellers insisted upon "machine gray" because it provided minimal distraction from the product's engineering and symbolized metallic qualities.[31]

The example of Sellers and a few other leading firms represented a distinct minority in American industrial design until the late nineteenth century, as most companies persisted in the use of ornament. Though some of these may have been manufacturers of inferior machines following Clegg's suggestion and hoping to snare buyers through meretricious ornament, this cannot be the whole explanation.[32] Works of undeniable merit often came in highly elaborated dress; the locomotive "America" offers one example. Furthermore, some of the taste for decoration unquestionably flowed from the consumer, and machine catalogues frequently offered optional painted details and decorative motifs. One is compelled to conclude that machine ornamentation was more than the refuge of the second-rate, that it ultimately sprang from far greater depths than the simple exigencies of marketing. It reflected profound aesthetic and social needs.

In our own time, most commentators have dwelt upon the hypocrisy of such design. Denouncing the engineers of this period as "sinners" and "sentimentalists," Lewis Mumford has protested: "In the act of recklessly deflowering the environment at large, they sought to expiate their failures by adding a few sprigs or posies to the new engines they were creating: they embellished their steam engines with Doric columns or partly concealed them behind Gothic tracery: they decorated the frames of their presses and their automatic machines with cast-iron arabesque. . . . Everywhere similar habits prevailed: the homage of hypocrisy to art." More recently, Marvin Fisher has contended that the elaborate artifice of nineteenth-century American machine design marked the suppression of a conflict of national values between reverence for nature and technological civilization. By embellishing and disguising machines so that they conformed to traditional notions of beauty, Fisher suggests, Americans attempted to make technology acceptable and harmonious with the environment, when in fact it profoundly threatened that environment.[33]

Such criticisms powerfully illuminate the underlying contradictions in nineteenth-century America's cultural response to technology. But analysis must not stop here. In concentrating on what Americans denied or ignored, one risks missing altogether the message of what they affirmed. Machine ornamentation represented an effort by engineers, manufacturers, and American society as a whole to assimilate the machine into republican civilization, to signify its honored status in the life of the nation, and to enhance its reputation as art. In the architectural embellishments, bright colors, and decorative motifs, one may see the conviction, at once defensive and assertive, that machinery belonged at the center of American culture.[34] Occasionally, one even discovers items which make explicit the technologists' boast that machinery constituted the true republican art form. A work such as the "Columbian" printing press (first made in 1813), with its claw-and-ball feet, caducei, and serpentine figures, is obviously anathema to functionalist taste. Nevertheless, it reflects an undeniable aesthetic impulse. The eagle atop the bar symbolically affirms the fruits of the union of technology and republicanism as he clutches, in addition to an olive branch and arrows, a cornucopia.[35] Mumford and Fisher would argue the essential evasiveness of such design, which distracts the viewer from the machine's function; and while this is true enough to modern taste, nineteenth-century Americans looked at technology with different eyes and a different aesthetic. With few exceptions they simply did not approach machines seeking pleasure in "precision, economy, slickness, severity, restriction to the essential"—Mumford's canons of machine art.[36] Rather, they derived from technology many of the same aesthetic values that they demanded of the fine arts. Far from anticipating Mumford's taste, they admired elegance, exoticism, and imagination as the passion for picturesque eclecticism in architecture and the decorative arts suggests. Instead of objects that made their appeal strictly in utilitarian terms, they valued works that brought the viewer the greatest associations, the most powerful emotional resonance. Approaching form as "implicit movement," they delighted in designing and viewing objects as expressive of human emotions. Paintings, buildings, furniture all characteristically surged with action to present a continuing drama of emotional metaphor.[37] These values were easily extended from the fine arts to machinery. Viewers responded to the explicit movement of machines at work as they did to the implicit movement of art, and they discovered in a factory or machine exhibition a gigantic animated theater of this expressive aesthetic.[38] When in addition these machines were brightly painted or garbed in Greek Revival or Gothic dress, their invitation to be viewed as art became almost irresistible. As works that satisfied the national passion for both beauty and utility, productive achievement and artistic expression, American machinery formed a major part of the aesthetic experience of the nineteenth century.

The machine aesthetic, then, ultimately extended far beyond characteristics of design: it was a mode of perception. Ornamental machinery appealed strongly to Americans' desire for emotional expression; but one should not ignore the important role played by machines apart from any ornamentation. To examine more closely this aspect of the machine aesthetic, let us study the public reaction to a machine without extraneous ornament: the great Corliss steam engine at the Philadelphia Centennial Exposition of 1876. A two-cylinder engine capable of developing up to 2500 horsepower, the Corliss engine supplied power for all the exhibits in Machinery Hall. It stood 39 feet tall, weighed about 680 tons, and was built at a cost of $200,000 by George H. Corliss of Providence, Rhode Island, who furnished it free to the exposition.[39] John Kouwenhoven has hailed the engine as a masterpiece of functional design; but pure and appealing as its lines are to modern eyes, this was not the source of its overwhelming popularity and acclaim in the nineteenth century. Instead, visitors to the Centennial were chiefly impressed by the engine as a powerful, indeed monumental, symbol of man's technological triumphs and a titanic form to inspire the romantic imagination. Detailed knowledge of mechanical functions was not only superfluous for these seekers of the technological sublime, but actually threatened to impair the free flow of associations. Machinery Hall, declared a writer for the *Atlantic Monthly*, "makes an extraordinary impression upon everybody, and probably those who understand nothing of what they see are more imaginatively affected than those who know all about valves and pistons."[40] Nineteenth-century engineering critics protested that the Corliss engine represented no important technical innovations and even suggested that with a few improvements its power and efficiency might be vastly increased.[41] But fairgoers did not approach the engine as an immaculate work of engineering, to be judged by its efficiency alone. Rather, in a way characteristic of popular reactions to powerful machinery in the nineteenth century, their descriptions frequently became incipient narratives in which, like some mythological creature, the Corliss engine was endowed with life and all its movements construed as gestures. The machine emerged as a kind of fabulous automaton—part animal, part machine, part god. In the melodrama which various visitors projected, the engine played the part of a legendary giant, whose stupendous brute force, they congratulated and titillated themselves, was harnessed and controlled by man. When President Grant and Dom Pedro, Emperor of Brazil, led the official proces-

Heroic technology: John Ferguson Weir's *The Gun Foundry*.

sion into Machinery Hall to start the engine at the opening of the Centennial, the audience thrilled to see how a slight human gesture could trigger such vast power. The description in the *Scientific American Supplement* is a classic of the genre. As the reporter presented the scene, the dignitaries entered "a concourse of genii" whose "mechanical forms crouched low along the nave," resting on "huge iron paws," "their hoppers yawning for victuals." Finally, "looking like pigmies," the procession arrived at the "Cyclopean" Corliss engine.

"Now, Mr. President," said Mr. Corliss. "Well," said the President quietly. "How shall I do it?" "Turn that little crank around six times." General Grant made a motion with his fingers inquiringly, "This way?" "Yes." In another half minute the screw was turned by General Grant, the colossal ma-

chine above him began to move, the miles of shafting along
the building began to revolve, several hundreds of steel and
iron organisms were set going.

"Strong men," the New York *Herald* reported, "were moved to
tears of joy."[42]

The power of the Corliss engine to fascinate and inspire as an
artistic spectacle received testimony from numerous and diverse
sources. "The first thing to do," a popular guidebook exhorted
fairgoers, "is to see the tremendous iron heart, whose energies are
pulsating around us." Paying tribute to the creative achievement of
George Corliss, the guide continued, "Poets see sublimity in the
ocean, the mountains, the everlasting heavens; in the tragic elements of passion, madness, fate; *we* see sublimity in that great fly-
wheel, those great walking-beams and cylinders, that crank-shaft,
and those connecting rods and piston-rods,—in the magnificent
totality of the great Corliss engine." Exclaimed a reporter for the
Nation, "Nothing could surpass the grace with which the giant
does his work. It is to me the highest embodiment of man's power,
and I could stand by the hour watching it." (Walt Whitman reportedly did precisely that.) Meanwhile, in her Malapropian dia-
lect, the humorous character "Josiah Allen's wife" enthused over
the inspirational force of the engine on behalf of common Centennial visitors: "Why, if there hadn't been another thing in the
hull buildin', that great 'Careless Enjun' alone, was enough to run
anybody's idees up into majestic heights and run 'em round and
round into lofty circles and spears of thought, they hadn't never
thought of runnin' into before." Her wondrous response was
confirmed by a description in one of the Centennial's numerous
unofficial catalogues. There the writer saluted the Corliss engine
as "that wonder of the modern era, that thing which needs
but the breath of life . . . to be a creation. . . . There it stands,
holding its place as a veritable king among machinery, so powerful
and yet so gentle, . . . without whose labor our efforts would be
small indeed; the breathing pulse, the soul of the machinery exhibition."[43]

This vital fusion of power and gentleness strongly impressed
William Dean Howells in his visit to the Centennial as editor-in-

chief of the *Atlantic Monthly*. In America's machine displays as a whole, rather than in her painting, sculpture, or literature, he ruefully confessed, he discovered the freest expression of "the national genius"; and despite his aesthetic predilection for chatty realism, he too felt sublimity in the presence of the Corliss engine. Although Howells protested that the engine's effect could not be easily conveyed in cold print, he began his description rhetorically insisting upon its majesty and power: "It rises loftily in the centre of the huge structure, an athlete of steel and iron with not a superfluous ounce of metal on it; the mighty walking-beams plunge their pistons downward, the enormous fly-wheel revolves with a hoarded power that makes all tremble, the hundred life-like details do their office with unerring intelligence." However, Howells shifted his tone to sound a pastoral note as he considered the almost effortless human control over this mighty engine, the garden, so to speak, in the machine: "In the midst of this ineffably strong mechanism is a chair where the engineer sits reading his newspaper, as in a peaceful bower. Now and then he lays down his paper and clambers up one of the stairways that cover the framework, and touches some irritated spot on the giant's body with a drop of oil, and goes down again and takes up his newspaper." Then, in a recurrent image in American descriptions of technology, Howells observed that the engineer resembled "some potent enchanter there, and this prodigious Afreet is his slave who could crush him past all semblance of humanity with his lightest touch." Having raised this threat, he immediately submerged it again, adding that in Machinery Hall one thought not of such perils but only of "the glorious triumphs of skill and invention."[44]

In analyzing similar passages, recent commentators have concentrated upon submerged anxieties in the presence of modern technology. Beneath the image of the machine as beneficent servant of mythological proportions, they have discovered a countervailing metaphor of the machine as monster, suddenly throwing off human control and unleashing its destructive force. This and related imagery, the argument runs, betray the profound uneasiness, the high emotional cost, with which Americans displaced the natural with a technological order.[45] Certainly a number of writers of the period, notably Herman Melville, consciously exploited the symbolic possibilities of this theme.[46] However, one must be extremely wary of extrapolating from these instances a similar, subconscious concern in the culture as a whole. The point is not an idle one; it involves the aesthetic and emotions with which nineteenth-century Americans perceived technology.

The basic difficulty lies in modern observers' remoteness from and consequent misunderstanding of the aesthetic of the technological sublime which is here at work. Although many critics have recognized the presence of the sublime in the ebullient, celebratory rhetoric of the period, they have neglected to consider in the same spirit the sense of awe and dread Americans discovered in the presence of machinery. Yet these had been key elements in the concept of the sublime as it developed in the course of the eighteenth century. As Edmund Burke declared in his *Enquiry* of 1757 on the sublime and beautiful, "Whatever is fitted in any sort to excite the ideas of pain, and danger, that is to say, whatever is in any sort terrible . . . is a source of the *sublime*, . . . the strongest emotion which the mind is capable of feeling." According to Burke, the sublime represented that delight which men have in "an idea of pain and danger, without being in such circumstances." Among the qualities and conditions which evoked these emotions were power, vastness, darkness, obscurity, suddenness, and noise. Burke's remarks concerning power have particular relevance here:

> Strength, violence, pain and terror, are ideas that rush in upon the mind together. Look at a man, or any other animal of prodigious strength, and what is your idea before reflection? Is it that this strength will be subservient to you, to your ease, to your pleasure, to your interest in any sense? No; the emotion you feel is, lest this enormous strength should be employed to the purposes of rapine and destruction.

Burke, of course, nowhere mentions machine technology as a source of the sublime; the closest he comes is to cite artillery as among those noises which awaken "a great and aweful sensation in the mind."[47] However, in the second half of the eighteenth century in England, George Robertson, Joseph Wright of Derby, and other artists soon broadened the field of the sublime beyond the natural landscape and discovered new sources of sublime emotion in industrial processes. By the mid-nineteenth century, both in

The "solemn demon face" of the machine: A Southern cotton press by night.

England and America, this aesthetic of the technological sublime had achieved a broad following, as the reactions to the Corliss engine dramatically illustrate. The desire to "see sublimity . . . in the magnificent totality of the great Corliss engine," to shiver with awe and dread before it, had become a popular passion. True, it may be in part that the sublime provided an aesthetic by which to express the ambivalent emotions that machine technology aroused. But even so, it was an aesthetic experience that people controlled, enjoyed, and sought again and again. Invocations of a steam engine's destructive power or descriptions of locomotives belching sparks and smoke, then, were not Freudian slips or subconscious tokens of anxiety. Rather these were images upon which the age placed a positive aesthetic value; they represent the conscious cultivation of the sublime.[48]

The way in which the technological sublime was at once stimulated and fused with a progressive vision may be illustrated by a consideration of the greatest American painting of an industrial scene in the nineteenth century, John Ferguson Weir's *The Gun Foundry* (Putnam County Historical Society, Cold Spring, New York) of 1866. Almost all of Weir's youthful education and experience seem to have combined to equip him to paint it. Born in 1841, Weir received early and thorough artistic training from his father, Robert Weir, himself a highly accomplished painter and drawing instructor at the United States Military Academy at West Point. Across the Hudson River at Cold Spring lived Gouverneur Kemble, an art patron and founder of the West Point Foundry, manufacturers of government cannon, who frequently entertained members of the Weir family at his elegant bachelor dinners. It was here at the foundry that J. F. Weir found the subject which occupied his best efforts as a young and ambitious painter. After brief military service in 1861, he launched his artistic career by preparing studies which would lead to *The Gun Foundry* and its companion piece, *Forging the Shaft* (original of 1867, destroyed by fire; replica, completed a decade later, in Metropolitan Museum of Art, New York). The casting of large guns and forging of shafts for side-wheeler steamships, "the primitive and massive appliance employed in this heavy work in the dingy shops, with the men sometimes stripped to the waist toiling vigorously," he later recalled, "appealed to my imagination with a significance beyond that of the merely picturesque."[49] Thus moved, he devoted to *The Gun Foundry* energy and passion appropriate to his subject. He spent weeks making the initial studies of the foundry's separate features. Then, in late 1864, re-examining his foundry sketches,

Weir was struck anew by "the virile activities of that busy place; the thought of it, in looking over the studies, seemed to take strong hold of the imagination for its human interest." Spreading his studies on the walls and floor of his studio one night, he began sketching a large charcoal cartoon of *The Gun Foundry* and worked until dawn. Nearly two more years and repeated visits to the foundry were required to realize this initial arrangement in the final work. His deliberate approach was rewarded by the encouragement and praise of some of America's most distinguished painters and, when the work was finished, widespread critical acclaim.[50]

Weir's fascination with the life of the foundry manifests itself clearly in the painting. He has chosen to depict the casting of a Parrott rifled cannon, invented by Robert Parrott, the current proprietor of the foundry and purchaser of the painting, at its most perilous moment. In the foreground a team of men assist in pouring molten iron from a huge caldron into a molding flask sunk in a deep pit. In the middle distance another group of workers grips the handles of the crane suspending the caldron. A lone figure tends the furnace in the distance. At right a soldier escorts a visiting party (Mr. and Mrs. Parrott, his brother-in-law Gouverneur Kemble, and other members of the Kemble family). The canvas was forty-eight by sixty-three inches, much larger than anything Weir had attempted before, and he composed his picture with a sense of vastness and monumentality capable of sustaining a still larger work. Human figures move in only a third of the area of the painting. They are enveloped by the industrial setting: the huge crane and caldron, the massive furnace-stack, the high rafters and brick walls. The glow of molten iron provides the main source of light, throwing strong shadows and casting the deeper recesses of the forge into dusk and obscurity. Still, the workers are in no danger of being subsumed by their surroundings. Weir has presented them as figures of immense strength and energy. With their sleeves rolled up or stripped to the waist, they reveal broad shoulders and powerful, bulging arms. Pushing and pulling in their work with one leg bent and the other straight in a continuous line with the back, they strike variations on what the art historian Kenneth

Clark has called the "heroic diagonal," a gesture that since the ancient Greeks has symbolized vigor and resolution.[51] While Weir exploits the dangerous aspect of his subject to provide dramatic tension, he nevertheless checks it by enlisting confidence in the workers' expertise. Animated and alert, totally absorbed in their task, they move about without a hint of panic. As Weir's treatment makes clear, they are to be regarded as industrial heroes in the epical struggle for the future of the Union.

This heroic aspect is central to the painting. *The Gun Foundry* followed a long and distinguished tradition of forge and foundry paintings dating back to the Renaissance, in many cases works whose themes were rooted in the mythology of Venus and Vulcan.[52] Though the mythological elements have outwardly receded in Weir's work, it is infused with their spirit. Behind Parrott's foundry still stands Vulcan's forge. Indeed, the painting implicitly suggests an analogy between the casting of this huge cannon for the defense of the Union and the forging of an immense shield by Vulcan and the Cyclopes for the protection of Aeneas and the Romans. Virgil's description in the *Aeneid* resonates in Weir's depiction:

> Rivers of metal flow, of brass
> And gold. In the huge furnace melts the steel,
> The creature of the fire. A mighty shield,
> Alone enough for all the Latins' spears,
> They forge; seven fold they make it, orb on orb.
> While some with bellows suck and force the air,
> Others plunge in the trough the hissing brass.
> Beneath the blows that fall the anvil rings.
> With mighty force alternately their arms
> They lift, each keeping stroke, while e'er they turn
> With tightly gripping tongs the hammered mass.[53]

This Virgilian parallel was immediately apprehended by a contemporary reviewer of Weir's picture, who compared the distant workmen to "Cyclop[e]s at their toil."[54] Within the realistic mode of the painting surges a sense of mythic grandeur that intensifies the sublime atmosphere while containing it within a positive symbolism. Thus the foundry visitors stand rapt and contemplative before the scene. Their dog leans back on his hind legs, his tail erect, staring at the spectacle. Even he knows he is in the presence of the sublime. The total effect is a compelling and ultimately reassuring tribute to the wonder of American technology.

While few nineteenth-century painters followed Weir's example and depicted the sublimities of American technology in oils, it became a favorite topic among popular illustrators. The pages of a leading illustrated journal, *Harper's Weekly*, for example, abound with pictures of blast furnaces and Bessemer converters, railroad stations and cotton-presses. Frequently these were portrayed at night, when their mystery and majesty appeared intensified.[55] The accompanying texts reinforced the sublimity of the scenes. Often these descriptions bore a highly "literary" character, indicating a calculated effort to apply a story to the visual image in order to give the reader a delicious shudder while exulting in the achievements of modern technology. J. O. Davidson's explanatory note to his engraving *Interior of a Southern Cotton Press by Night* offers one instance:

> Beneath the converging rays of electric lamps and reflectors a most weird effect is produced, for the machine assumes the aspect of a grand and solemn demon face, strangely human, recalling the famed genii of the *Arabian Nights*. Beside it are the furnaces, whose open doors glow with the fires supplying the vitality of the giant, while about them flit the half-naked forms of the firemen—attendant demons of this monster.

Or consider this description of a gun press at Bethlehem, Pennsylvania, by the painter and early historian of American art, S. G. W. Benjamin: "The press comes down with the remorseless certainty of fate, noiselessly but as terrible as the knife in Poe's famous and horrible story of 'The Pit and the Pendulum.' When the press strikes the molten iron, flakes of fire are forced up that fly forty or fifty feet, and the by-standers must keep clear of the burning missiles." As the passage continues, this Gothic thriller atmosphere gives way to frank admiration of the machine's achievement: "This press bears down with the stupendous power of several thousand tons. If one were to concentrate the weight of a modern iron-clad ship of war, with guns and equipment and crew, it would scarcely be more than the force concentrated in this most tremendous engine yet constructed by man."[56]

Perhaps the most common vehicle of the technological sublime in the nineteenth century was the railroad. Viewers not only thrilled to the elaborate ornamentation of locomotives such as the "America," which reached its height around midcentury; they were still more fascinated by the sight of the railroad in motion. Universally accessible, a rushing train possessed almost all the Burkean attributes and symbolized the beneficent new technological order. Observed at close range, particularly at night, it possessed an irresistible appeal. In a diary entry of 1839 the urbane New Yorker George Templeton Strong enthused over the excitement of a large train and the technological progress it represented. "Just imagine," he went on to add, "such a concern rushing unexpectedly by a stranger to the invention on a dark night, whizzing and rattling and panting, with its fiery furnace gleaming in front, its chimney vomiting fiery smoke above, and its long train of cars rushing along behind like the body and tail of a gigantic dragon— or the d——l himself—and all darting forward at the rate of twenty miles an hour. Whew!" A reporter for *Scientific American* agreed. "One of the grandest sights in the world," he exclaimed, "is a locomotive with its huge train dashing along in full flight. To stand at night at the sight [*sic*] of a railroad, when a large train is rushing along at the rate of 30 miles per hour, affords a sight both sublime and terrific."[57] To these voices Walt Whitman added his song to a locomotive on a dark winter afternoon:

> Thy black cylindric body, golden brass and silvery steel,
> Thy ponderous side-bars, parallel and connecting rods, gyrating, shuttling at thy sides,
> Thy metrical, now swelling pant and roar, now tapering in the distance,
> Thy great protruding head-light fix'd in front,
> Thy long, pale, floating vapor-pennants, tinged with delicate purple,
> The dense and murky clouds out-belching from thy smoke-stack,

A bucolic view of the railroad: George Inness's *The Lackawanna Valley*.

Thy knitted frame, thy springs and valves, the tremulous twin-
    kle of thy wheels,
Thy train of cars behind, obedient, merrily following,
Through gale or calm, now swift, now slack, yet steadily
    careering;
Type of the modern—emblem of motion and power—pulse of
    the continent,
For once come serve the Muse and merge in verse, even as here
    I see thee. . . .

For Whitman the "pant and roar," the thick belches of smoke, no
less than the shiny fittings and picturesque cars, formed an essential
part of the railroad's artistic fascination and made it the cogent
"emblem of motion and power" it was. These sublime emotions
stirred even the very young. At the time Whitman wrote this
poem in 1876, the future painter Lyonel Feininger was already
forming strong impressions of locomotives, "half terrifying and

wholly fascinating." At the age of five, he later recalled, he "already drew, from memory, dozens of trains, ... the black locos of the [New York Central] with 'diamond' smokestacks, and the locomotives of the [New York, New Haven and Hartford Railroad] with elegant straight smokestacks painted, like the driving wheels, a bright vermillion red, and oh, the brass bands about the boiler and the fancy steam domes of polished brass."[58]

To many historians nineteenth-century Americans' refusal to recognize a moral and aesthetic conflict between this passionate embrace of the machine and their professed love of nature appears maddeningly perverse; yet the fact of that refusal remains. Emerson, Thoreau, and others might call for a controlling poetic vision that would reconcile the claims of nature and civilization; but the great bulk of Americans believed that the progress of American civilization contained its own controls. To be both Nature's nation and a rapidly developing industrial power was to them not a contradiction, but the fulfillment of America's destiny. The natural landscape and modern technology provided the resources to complement one another. One might therefore clothe both in visions of sublimity.

This is, to be sure, a very different state of mind and aesthetic than the pastoralism Leo Marx discusses so well in *The Machine in the Garden*. If technology functioned as a "counterforce" in many of the literary reveries of the century,[59] much more frequently in the general culture technology represented both socially and aesthetically a welcome source of excitement, an addition but not a fundamental disruption of the natural order. To explore fully a subject as large as this one would require a book in itself. Some aesthetic responses to the railroad's appearance in the landscape, however, may be briefly considered. A few Americans, like Daniel Webster, indulgently cited the claims of the peaceful landscape against the railroad's invasion only to dismiss them;[60] but Edward Everett was more typical in arguing that the railroad in fact facilitated the contemplation of nature. Speaking in Boston in 1851, he remembered that a similar issue had arisen in England when a railroad was proposed through the lake country, with no less a figure than William Wordsworth opposing the project. Everett commended Wordsworth's motives, but contended that in this

A sublime view of the railroad, as seen by Currier & Ives.

matter they were quite misdirected. "The quiet of a few spots may be disturbed by a railroad," he acknowledged, "but a hundred quiet spots are rendered accessible. The bustle of the station-house may take the place of the druidical silence of some shady dell; but gracious heavens! sir, how many of those verdant cathedral arches, entwined by the hand of God in our pathless woods, are opened for the first time since the creation of the world, to the grateful worship of man by these means of communication!"[61]

Occasionally, railroad companies organized special excursions to advertise their roads in a way which supported Everett's argument. In 1858 the Baltimore and Ohio invited a party of about forty painters, photographers, writers, and others associated with the arts to a five-day railroad excursion. The group, which included John F. Kensett, Asher B. Durand, Thomas Rossiter, and Nathaniel Parker Willis, traveled from Baltimore to Wheeling, Virginia, and back, enjoying both the dramatic scenery and a holiday camaraderie. They returned inspired by picturesque and sublime views, but also impressed by the railroad's technological triumphs, which at least once prompted them to a round of cheers.[62] In 1872 a group of guests of the Northern Central Railway Company explored the countryside from Washington, D.C., to Niagara Falls in a similar vein. Although their chronicler indulged in some jibes at the railroad's expense, the whole party arrived at Niagara (itself America's most celebrated scene of natural sublimity) much impressed with the road they had traveled and its recreational and commercial potential. The group's poet spoke for them all in his verses commemorating a night ride together. Unmemorable as poetry, his lines are nonetheless valuable as cultural history. Celebrating the railroad as the greatest of all forms of transportation, the poem enthusiastically describes a locomotive snorting steam and terrifying the earth with its roar.

> Crossing long and thread-like bridges,
> Spanning streams, and cleaving ridges,
> Sweeping over broad green meadows,
> That in starless darkness lay—
> How the engine rocks and clatters,
> Showers of fire around it scatters,
> While its blazing eye outpeering
> Looks for perils in the way.

The poem concludes in a chant of delirious joy:

> Oh! what wildness! oh! what gladness!
> Oh! what joy akin to madness!
> Oh! what reckless feeling raises
> Us to-day beyond the stars!
> What to us all human ant-hills,
> Fame, fools sigh for, land that man tills,
> In the swinging and the clattering
> And the rattling of the cars?[63]

Others, to be sure, attempted to view the railroad's relationship to the landscape in a more pastoral mode. The outstanding example of this effort is George Inness's *The Lackawanna Valley* (formerly known as *The First Roundhouse of the D.L. & W.R.R. at Scranton*, National Gallery of Art, Washington, D.C.) of 1855. Like the two excursions discussed above, the painting sprang from a railroad company's promotional attempt. The Delaware, Lackawanna and Western Railroad commissioned Inness to paint the company's new roundhouse at Scranton, Pennsylvania, as an advertisement. Although he undertook the job somewhat reluctantly, Inness rose to the artistic challenge and achieved a startlingly successful integration of the railroad with the bucolic landscape. The industrial buildings in the background nestle gently between the far hills and the groves of trees in the middle distance. The train itself puffs through newly cleared farmland along a gracefully undulating track. The solitary, contemplative country boy, reclining dreamily in the foreground, provides a viewer with a final cue to his response. Inness's painting celebrates less the Lackawanna Railroad's technological triumphs than an Arcadian mood still firmly rooted in agrarian America. Significantly, however, this depiction did not serve the railroad company's purposes, and they declared the work unsatisfactory. They insisted Inness show all four trains of the road in the painting, and that the letters "D.L. and W." appear on a locomotive. Clearly, if Inness wished to reduce their line to a picturesque feature in the landscape, they would have none of it. Needing the money, the artist finally agreed and made the necessary changes. He received seventy-five dollars. The company later sold the painting, and over thirty years later, Inness recovered it in a curiosity shop in Mexico City.[64]

The leading American lithographic publishers, Currier & Ives, who produced several advertisements for railroad companies, were far more successful in expressing the popular conception of the locomotive's place in the landscape. Of their numerous railroad lithographs, some, such as *The Night Express: The Start* (undated) or *Night Scene at an American Railway Junction* (1876), depict the train in its most sublime aspect. The latter portrays a railroad station at night in a symphony of red and black. Smoke pours from the engines' smokestacks and steam hisses from their brakes as passengers and workmen scurry among the cars. The whole scene surges with movement, dominated by the arriving trains themselves. Other Currier & Ives prints, however, present distant views of trains chugging through green valleys and dramatic scenery. *Lookout Mountain, Tennessee and the Chattanooga Railroad* (1866) stands as perhaps the finest example of this kind of lithograph among Currier & Ives's large folios and *The Great West*

(1870) and *Through to the Pacific* (1871) among the small folios. Although in all these prints the locomotive has a more toylike aspect than in the close-up views, the artists nevertheless have not attempted to merge the train with the landscape. Rather, as in many of the "naïve" paintings of the same period, the train *animates* the scene.[65] Each of these pictures reveals a sense of pleasure and triumph at the spectacle of the railroad moving easily through formerly difficult terrain, which recalls Whitman's lines from "Passage to India":

> I see over my own continent the Pacific railroad surmounting
>   every barrier,
> I see continual trains of cars winding along the Platte carrying
>   freight and passengers,
> I hear the locomotives rushing and roaring, and the shrill
>   steam-whistle,
> I hear the echoes reverberate through the grandest scenery in
>   the world. . . .[66]

Currier & Ives's most ambitious statement of the relationship of the railroad to the landscape was the lithograph *Across the Continent* of 1868. Anticipating the completion of the transcontinental railroad the next year as Whitman did in "Passage to India," it is subscribed "Westward the Course of Empire Takes its Way."[67] James M. Ives and one of the firm's leading artists, Mrs. Frances Flora Bond Palmer, began work on the picture as early as 1862. Two rough sketches in her hand bear that date, along with suggested improvements by Ives. For the final composition Mrs. Palmer combined the two sketches; Ives chose the title and probably drew the figures.[68] In the completed work a train bound from New York to San Francisco streaks across the prairie, dividing the picture in a sharp diagonal into the realms of nature and civilization. On the right, before a peaceful river and majestic mountains, two Indians on horseback watch amazed as the locomotive rushes by, forcing them literally to eat its dust. They, as much as the bison in the distance, represent the savage state of the American wilderness, which, the picture makes clear, is to be superseded by a "higher" civilization of American enterprise. To the left a frontier village epitomizes from its flowers to

The engine of republican civilization, another vision by Currier & Ives.

the log-cabin church the effort of hearty, independent pioneers to nurture and civilize the land. Children from a clearly designated "PUBLIC SCHOOL" run out and cheer the train. Woodcutters clearing the land pause contentedly. By the side of the tracks men are erecting telegraph poles, and settlers in covered wagons are starting across the plains. Here is no attempt to integrate the railroad into the natural landscape as in Inness's painting. On the contrary, the railroad is seen as revolutionizing the prairie. It is at once the agent and symbol of America's republican civilization, extending the empire of liberty, taming the wilderness, and turning it into a garden.[69] And the viewer is intended to witness this transformation not from a contemplative, pastoral perspective as in *The Lacka-wanna Valley*, but marching along in the sweep of civilization the railroad brings, positively cheering it on.

Thus Emerson spoke acutely when he observed in his journal, "The Railroad is that work of art which agitates & drives mad the whole people; as music, sculpture, & picture have done on their great days respectively."[70] For a society that sought dynamic, productive, and expressive images to embody its underlying principles, the fine arts alone were not sufficient. Though Emerson himself regarded the railroad as a limited art unless inspired by a vision of noble ends, the broader public discovered in it and in machine technology generally the combination of powerful emotion, utility, and moral purpose which they demanded for republican art. In this respect, it mattered little that most American writers and painters continued to adopt an attitude of studied indifference toward technology in their works. Despite their hostility, which of course was often returned in full measure by technologists, not only was technology adopted as a significant subject for art, particularly in the popular arts; it established a position as a popular art in its own right. The form and, more important, the aesthetic interpretation of American machinery thus developed in response to republican ideology, which technology itself had helped to shape.

5

Technology
and Utopia

IN THE closing decades of the nineteenth century the tension between America's commitment to republicanism and her rapid technological growth reached a crisis. With dizzying swiftness the United States came of age as a technological society, and the transition affected virtually every sector of American life. A revolution in transportation and communications, led by the extraordinary expansion of the railroad, converted what had been a society of "island communities" into a centralized nation with vital links throughout the world.[1] The creation of new mass markets stimulated a transformation in all facets of industry. The steel, petroleum, and electrical industries quickly grew from infancy to gigantic proportions. Rapid implementation of new machines, interchangeable parts, and standardized processes allowed older industries to expand radically as well. From the fourth-ranked position among manufacturing nations of the world as of 1860, the United States pulled into a decisive lead. By 1894 the value of her manufactured products nearly equaled that of Great Britain, France, and Germany combined.[2] The social and cultural dislocations that attended this great technological transformation were immense. On the celestial railroad of American development, many found the journey disorienting, the roadbed rough, the milestones blurred, and the stations unfamiliar. Some were shut in boxcars or even condemned to ride the rods. Yet for a time their voices were muted as those who reclined in Pullman cars cheered on the nation's progress. Celebrating the stupendous achievements of America's "triumphant democracy," the steel magnate and Scottish immigrant Andrew Carnegie thumbed his nose at European rivals: "The old nations of the earth creep on at a snail's pace; the Republic thunders past with the rush of the express."[3]

An era of unprecedented production, the late nineteenth century was also a brilliant age of invention. A crude index of its achievement lies in the sheer traffic through the U.S. Patent Office. The record of 23,000 patents issued during the decade of the

1850s (which quadrupled the amount of the previous ten years) was approximated if not excelled during every single *year* from 1882 on. Bestriding the field of invention like a colossus was the great popular hero of the period, Thomas Edison. The inventor of the multiplex telegraph, an improved telephone transmitter, the mimeograph machine, the phonograph, the microphone, the motion picture, in addition to his celebrated development of a practical incandescent lamp, and countless other contributions in the field of electrical engineering, Edison literally made a business of invention. At a time when technological innovations were becoming increasingly complex and dependent upon science inaccessible to the layman, "the wizard of Menlo Park" provided a reassuring image of a self-taught, self-made man who could master the forces of nature and convert them to practical purposes. In the popular accounts of the time he became almost an allegorical representation of republican technology, both literally and figuratively dispelling "night with its darkness . . . from the arena of civilization."[4]

So firmly and immediately did the nation embrace Edison's and other technological developments that an American in 1900 looking back over the course of the last century could hardly imagine life without them. Writing in that year, a historian of invention, Edward W. Byrn, attempted to dramatize what moving backward in time a hundred years would entail. Leaving the present in "a luxurious palace car behind a magnificent locomotive, traveling on steel rails, at sixty miles an hour," one neared the beginning of the nineteenth century in "a rickety, rumbling, dusty stagecoach." Outside, in a ghostly panorama, the milestones of technological progress receded: not only telephone, phonograph, camera, electric railways, and electric lights, but as the journey progressed, telegraph, sewing machine, reaper, thresher, and India rubber goods. Americans lost the steam-powered printing press, woodworking machinery, gas engines, elevators, barbed wire, time locks, oil and gas wells, and ice machines. Faster and faster flowed the motley procession of inventions until Byrn's catalogue came in a rush:

> We lose air engines, stem-winding watches, cash-registers and cash-carriers, the great suspension bridges, and tunnels, the

Suez Canal, iron frame buildings, monitors and heavy iron-clads, revolvers, torpedoes, magazine guns and Gatling guns, linotype machines, all practical typewriters, all pasteurizing, knowledge of microbes or disease germs, and sanitary plumbing, water-gas, soda water fountains, air brakes, coal-tar dyes and medicines, nitro-glycerine, dynamite and guncotton, dynamo electric machines, aluminum ware, electric locomotives, Bessemer steel with its wonderful developments, ocean cables, enameled iron ware, Welsbach gas burners, electric storage batteries, the cigarette machine, hydraulic dredges, the roller mills, middlings purifiers and patent-process flour, tin can machines, car couplings, compressed air drills, sleeping cars, the dynamite gun, the McKay shoe machine, the circular knitting machine, the Jacquard loom, wood pulp for paper, fire alarms, the use of anaesthetics in surgery, oleomargarine, street sweepers, Artesian wells, friction matches, steam hammers, electroplating, nail machines, false teeth, artificial limbs and eyes, the spectroscope, the Kinetescope or moving pictures, acetylene gas, X-ray apparatus, horseless carriages, and—but enough!

Byrn shook himself and his readers from the thought as from a nightmare. To regress into the "appalling void" of the past, mired in ignorance and superstition, was unendurable. Far pleasanter was it to contemplate recent achievements and to anticipate future triumphs of technology, confident that no dream could be too bold.[5]

Byrn reflected the dominant popular conception of history as a steadily progressive record in which increased knowledge, technological development, and political liberty marched hand in hand. The course of technology was now bringing in sight a world civilization of reduced labor and enriched leisure, health and longevity, abundance, peace, and human brotherhood. Modern technology was giving birth to a "new epoch" of civilization that would penetrate around the globe, banishing ignorance, and eradicating "the savage and barbarous tribes" of the past. The vestiges of the primitive order would soon cease to function or even exist in the larger world and be confined to museums and academic studies.[6] By this view, technology was fast making Thomas More's famous pun in coining the word "utopia" an anachronism. Eutopia, the good place, it certainly was, but the qualification that it was also outopia, no place, no longer applied. Utopia lay precisely where many Americans had always contended, in the future, and every new invention attested to its existence and impending realization.[7] Every day America moved closer to its practical fulfillment.

If one looked from patent office and production records to the condition of American society in the late nineteenth century, however, the impediments to utopia appeared immense. Technology had been embraced as a principle of order and preserver of union, the harbinger of peace and guardian of prosperity. But the overriding paradox of the age was the coexistence of technological progress and social chaos. The industrial economy, despite its extraordinary growth, was hardly a smoothly running engine but an erratic and dangerous machine, capable of great bursts of activity, then inexplicable slumps. The first of these, the Panic of 1873, spurred a groundswell of industrial and agrarian discontent, climaxed by a violent nationwide railroad strike in 1877. Then, as middle-class fears of a revolutionary uprising subsided in the early 1880s, another massive railroad strike in 1885 and Chicago's Haymarket riot of 1886 ushered in twelve years of almost unrelieved social crisis, reaching pitches of despair during the depression of 1893–97. The attempt of George Pullman, the sleeping car king, to create a model factory town in Pullman, Illinois, while retaining control and reaping a profit over its operations, received sharper rebuke in the bitter strike of 1894 than Lowell ever did. The shock of rapid urbanization profoundly transformed the pattern of American life and compounded the problem of cultural division. As showcases of the new technology, America's great cities dazzled visitors, but they frequently appalled them as well with their vivid contrasts of wealth and condition. Neo-Renaissance palaces and rat-infested tenements, society hostesses and "fallen women," the crazy-quilt of ethnic differences all became clichés in the description of city life, and the image of a smoldering volcano a master-metaphor. Often observers darkly alluded to the Biblical parable of Dives and Lazarus, in which the rich man is damned and the poor saved. Despite the din of cheers on behalf of technology as the great social unifier and horn of plenty, a succession of titles such as Henry George's *Progress and Poverty* (1879), Jacob Riis's

How the Other Half Lives (1890), and Henry Demarest Lloyd's
Wealth Against Commonwealth (1894) pointed to fundamental
economic rifts in American society based upon inadequate distribution of the fruits of technology. "The march of invention," wrote
George, "has clothed mankind with powers of which a century
ago the boldest imagination could not have dreamed. But in factories where labor-saving machinery has reached its most wonderful development, little children are at work; wherever the new
forces are anything like fully utilized, large classes are maintained
by charity or live on the verge of recourse to it. . . . The promised
land flies before us like the mirage."[8] Though their nostrums varied, George, Riis, Lloyd, and other reformers agreed that industrial
America had strayed dangerously from both republican principles
and Christian ethics. Unless social justice was speedily granted,
they warned their readers, the oppressed might seek redress of
their own through violent revolution.

While these reformers won an attentive audience in the late
nineteenth century, a small but significant minority of American
intellectuals abandoned progressive visions and surrendered to
fin-de-siècle melancholy and sense of decadence. Imaginatively re-
treating to a golden age in the past, they discovered both a refuge
from the present and a position from which to criticize their own
era. Harvard professor of fine arts Charles Eliot Norton, Henry
Adams, and the architect Ralph Adams Cram, for example, taught
two generations of New Englanders to revere the Middle Ages
as a civilization of social unity, moral idealism, energy, beauty—
all the values they found lacking in industrial America. In the recent course of history they saw not progress but decline. " 'To live
for the future,' as we are told to do," Norton complained, "is to
live on the windiest and least nourishing of diets." His criticisms
of "this degenerate and unlovely age" in his Harvard lectures became legendary. Encouraged by Norton's example, in the 1890s
Cram and his self-consciously aesthetic circle went so far in their
medievalism as to repudiate republican principles altogether and
embrace monarchism. Henry Adams's repudiation was still more
cosmic. In the late nineteenth and early twentieth centuries he increasingly insisted that democracy, society, indeed, the universe
as a whole were all undergoing a steady process of degradation

and dissipation in accordance with the laws of science. Behind illusions of progress and order lay the reality of chaos and decay which man was powerless to alter, so that history itself, Adams grimly insisted, would have to be redefined as "the science of human degradation."[9]

A more widely shared retreat from technological America was expressed in the pervasive nostalgia for the homogeneous, preindustrial village culture of the early nineteenth century. The disorienting effect of rapid social and technological change—what Alvin Toffler has called "future shock"—reached epidemic proportions in the 1880s and '90s.[10] In compensation, the past assumed new importance as an emotional center of stability and security. Opposing the centralizing tendency of modern transportation and communications, regional literatures sprang up throughout the country which evoked the lost youth and vanished way of life of a generation effaced, in Sarah Orne Jewett's phrase, by "the destroying left hand of progress."[11] The same theme figured prominently in the personal reminiscences of the period. Henry Adams dated quite precisely the eclipse of the world into which he was born: it was May 1844, with "the opening of the Boston and Albany Railroad; the appearance of the first Cunard steamers in the bay; and the telegraphic messages that carried from Baltimore to Washington the news that Henry Clay and James K. Polk were nominated for the Presidency."[12] When Henry James returned to the United States for a visit in 1904 after more than twenty years abroad, he too found himself "amputated of half my history" by the incursions of technological society. He discovered himself a stranger in his native land, his New York boyhood home gone, the old sights of the city compromised or completely obliterated by the new industrial landscape, the tempo increased, the commercialism rampant. Traveling to Washington Irving's house, Sunnyside, in Tarrytown, New York, James sensed that the very railroad on which he rode hopelessly compromised his literary pilgrimage. The train symbolized "the quickened pace, the heightened fever" that cut the modern world irretrievably from Irving's gracefully indolent time and condemned the tantalized James to grasp for "the last faint echo of a felicity forever gone."[13]

Not only the physical environment, then, but the emotional

texture of the world of the generation born before the Civil War was altered drastically in the late nineteenth century. From a shelter of the pleasure principle they had fallen into an anxious world of time-consciousness. Lewis Mumford has argued that the clock, which induces a mechanical concept of time abstracted from the natural order, is the key machine of the industrial era; and we have already seen its importance in the establishment of factory discipline in the case of Lowell.[14] In *Walden* Thoreau had noted how the railroad was whirling Americans' "pastoral life ... past and away" and manifestly altering their sense of time: "Have not men improved somewhat in punctuality since the railroad was invented? Do they not talk and think faster in the depot than they did in the stage-office?"[15] In the last quarter of the nineteenth century the technological order strengthened time regimentation. Thoreau's observation proved prophetic, as a system of standard time was instituted in 1883 to simplify and coordinate railroad timetables, sweeping away a hodgepodge of more than fifty different local times.[16] By the early 1890s time clocks were being introduced into offices and factories, and Frederick W. Taylor, the father of "scientific management," was conducting his famous time-motion studies, which broke every job down to its principal components to determine the most efficient pattern for each. So well did Americans internalize the exacting time requirements of their age that George M. Beard, a pioneer in the field of psychosomatic medicine, listed the perfection of clocks and invention of watches high among the causes of American nervousness. In his view, technological society in general exacted a stiff price for its comforts and conveniences in the historically unprecedented strain it placed upon the nervous system. "Modern nervousness," Beard concluded, "is the cry of the system struggling with its environment."[17]

This sense of contradiction between inherited values and sudden change, between technological progress and social discontent, between republican principles and the new industrial order welled up powerfully in the late nineteenth century. Frustrated in society at large, the desire for synthesis of technology and republicanism expressed itself most fully in utopian literature. In the late 1880s and 1890s over 150 utopian and dystopian novels were published in America, many of them centering on the crisis of American republicanism in a technological age.[18] Amid this deluge of utopian literature, four books emerge prominently as the most imaginative and searching considerations of this crisis: Edward Bellamy's *Looking Backward* (1888), Mark Twain's *A Connecticut Yankee in King Arthur's Court* (1889), Ignatius Donnelly's *Caesar's Column* (1890), and William Dean Howells's *A Traveler from Altruria* (1894).[19] Representing four different types of the genre, together these novels provide an unfolding dialogue on the problems of integrating technology within a republican order. In their own distinctive ways each assesses the results of the United States' first century of political, social, and technological development, the meaning and relevance of egalitarian ideals in a technological society, and the possibility of achieving a republican utopia. Thus, these works spoke compellingly to a large readership profoundly agitated by the trial of American society and values. *Looking Backward* and *Caesar's Column* were two of the outstanding best-sellers of their day; and while neither *A Connecticut Yankee* nor *A Traveler from Altruria* could rival them in circulation, as books by two of the most popular and respected authors of the period, they reached a substantial and influential audience.

The novel provided these and other authors with a popular forum to communicate their ideas; but the impulse to utopian literature sprang from deeper sources as well. In a confused and apocalyptic time the form of the utopian novel offered a mode of interpreting social experience. Such fiction exposed the contradictions of contemporary American life, either by contrasting them with an imaginary unified social order, or by extrapolating the chaos that would result if American society continued to develop along existing lines. The novel provided writers with a flexible medium in which they could render ideas in imaginative terms instead of merely stating them as abstractions. Both the reductive and creative demands of fiction enabled authors to articulate their social visions with greater clarity than in discursive writing. Narrative structure and the conventions of plot provided mechanisms of ordering the confusing flux of social change and of charting its direction by relating it to a fictive beginning and end.[20] Utopia in these novels thus functioned as "a speculative myth," a mode of projecting a vision of social possibilities at a time when Americans felt themselves at a turning point in history, but desperately in need of a sense of direction. The calendar itself appeared to invite both millennialist and catastrophic projections that the end of the century would mean the end of an era.[21] While popular historical romances in the late nineteenth century, such as *The Prisoner of Zenda* and *When Knighthood Was in Flower*, offered their readership a retreat to a mythic past, utopian writers generally cast their readers' vision forward to a reconciliation of the anarchic forces of the present, ultimately a release from history itself, in a mythic future. With some significant exceptions, these utopian authors departed from the traditional location of utopia at a *physical* remove from their own society—in the manner of Plato, Bacon, and More—and located it instead at a *temporal* remove, projecting a vision of technological society that might emerge out of the turbulent events of contemporary America.

However, by no means were all of these visions unqualified affirmations. A recurrent theme in much of this utopian fiction concerned the possibility of social breakdown or catastrophe if American technology and society were not harmonized and controlled. "The imagination of disaster" frequently colored these works, whether the ultimate conclusion was positive (utopian) or negative (dystopian), and the forms which this imagination took illuminated and often criticized the dominant assumptions of the progressive view of American technological development.[22] Both utopian and dystopian works pointed to the danger of barbarism at the heart of America's vaunted civilization rising out of the gulf between the nation's abundant resources and her shocking social inequities. In a society whose republican purposes had been obscured or corrupted, these writers emphasized, technology itself might serve as an instrument not of liberty but of repression, not order but chaos, not creation but destruction. The hopeful vision of an integrated technological republic struggled against the dreadful anticipation of technological tyranny and holocaust.

Far and away the most popular and influential of all utopian novels of the period was *Looking Backward*. Edward Bellamy began the book in late 1886 and sent the completed manuscript to the publishers the following August. Published in January 1888, *Looking Backward* enjoyed only moderate sales of 10,000 copies during its first year. In 1889, however, sales climbed to over 125,000, and during the 1890s the book became an international best-seller and the stimulus for a host of volumes that extended, qualified, or repudiated Bellamy's ideas.[23] The work even inspired a short-lived reform movement, the Nationalists, a motley assortment of amateur social theorists, socialists of various stripes, clergymen, prohibitionists, feminists, army veterans, Theosophists, and litterateurs. By 1890 the number of Nationalist clubs had climbed to some five hundred across the country, all eagerly discussing how best to implement Bellamy's vision, though by the mid-1890s their strength waned considerably.[24]

Bellamy propounded the technocratic utopia of *Looking Backward* in response to what he saw as the subversion of republican institutions by the growth of industrial capitalism. A journalist and novelist before his book's success launched him into reform politics, he lived almost his entire life in the mill town and industrial cousin of Lowell, Chicopee Falls, Massachusetts. Like numerous other nineteenth-century Americans, he first perceived "the inferno of poverty beneath our civilization" only when he traveled to Europe in 1868 at the age of eighteen. Upon his return he discovered "I had now no difficulty in recognizing in America, and even in my own comparatively prosperous village, the same conditions in course of progressive development."[25] Beginning in the early 1870s in editorials in the Springfield *Daily Union*, Bellamy pointed to the alarming concentration of wealth and power in the hands of a few and the ruthlessness with which they exercised control. The spectacle of deformed, emaciated, ignorant factory children in the local mills, a sight Americans had at one time thought impossible in their own country, shocked him profoundly. Their presence, he warned, offered undeniable proof that "a great wrong exists somewhere among us which is inflicting a vast amount of barbarity, a positive cruelty of monstrous proportion, upon these children and others like them in New England." As yet, however, Bellamy could suggest only piecemeal reforms, such as the establishment of half-time schools for child laborers.[26]

The root of this new social barbarism, Bellamy believed, lay in the rise of a new industrial aristocracy rather than in technology itself. He found the grossly inequitable distribution of the fruits of technology all the more intolerable because of his own fascination with new mechanical marvels and their potential social benefits. In his short story "With the Eyes Shut," for example, he joined in popular speculation on the possible social impact of Edison's phonograph.[27] Edison himself had predicted that his invention would revolutionize existing forms of letter-writing, dictation, books, clocks, music, and hundreds of other practices, disseminating and preserving new worlds of expressive sound for people now bound by cold symbols and print.[28] Bellamy imagined a society in which magazines, newspapers, books, and letters were all recorded on phonographs, so that for the general public reading and writing would have become skills as dead as Greek. As Bellamy depicted it, this technological transformation standardized and mechanized potentially every facet of experience. An inventor even has devised a greeting machine for clerics and politicians; a visitor need only press various buttons describing his situation to receive a hearty programmed response. Similarly, the phonograph has been married to the clock to extend time-discipline and increase social control. Equipped with portable timepieces, individuals are now obliged to follow orders from their family or superiors literally like clockwork. As the story ends, the narrator discovers that he has been pleasantly dreaming on a train and that these inventions still remain to be developed.

In *Looking Backward* Bellamy brought together his distress at the growing plutocratic control of American institutions and his vision of a truly egalitarian technological republic into a unified work contrasting contemporary American society with his utopian alternative. Set in the year 2000, the book cleverly begins with a sham preface that premises the attainment of Bellamy's utopian society and proposes in the guise of historical romance to "look backward" at the contrasts between the nineteenth and twentieth centuries. The narrator of this fiction, Julian West, presents himself as uniquely qualified to compare the old order and the new. For, as he explains, he himself was once a wealthy and idle aristocrat living in late nineteenth-century Boston. As a member of the privileged class, he felt no concern with the misery of the mass of society, only resentment at the workers for their paralyzing strikes which delayed construction of his new house and marriage to his fiancée, Edith Bartlett. Nonetheless, West suffered from insomnia, a symptom of the nervous spirit of the age and the troubled soul of unregenerate man. In order to sleep, he habitually retreated to "the silence of the tomb" of a subterranean chamber, symbolic of his isolation and spiritual death.[29] One night, according to his occasional practice, West called a mesmerist to this vault to help him fall asleep. When he awoke, however, it was no longer 1887 but 2000.

To catapult his hero suddenly into the future in this way afforded Bellamy not only a convenient device to contrast his contemporary and utopian societies; in its extreme violence it provided a powerful metaphor for the painful sense of dislocation due to rapid change that characterized the late nineteenth century. Thus West is seized with terror as he realizes he is suspended in an alien world a century beyond anything he has ever known: "There are no words for the mental torture I endured during this helpless, eyeless groping for myself in a boundless void." Struggling to regain a sense of personal identity and continuity, he falls on a couch and fights for sanity: "In my mind, all had broken loose, habits of feeling, associations of thought, ideas of persons and things, all had dissolved and lost coherence and were seething together in apparently irretrievable chaos. There were no rallying points, nothing was left stable." Here is a dramatic rendering of the condition Henry Adams described in his *Education*: "He saw before him a world so changed as to be beyond connection with the past," and like an earthworm, he "twisted about, in vain, to recover his starting-point."[30] In his description of West's discovery that he has awakened into a world totally new, Bellamy at once emotionally validated his readers' own sense of historical disorientation and presented a vision of the culmination of history in a stable and beneficent social order.

For West's revival is the beginning of his resurrection. By the year 2000 society has realized in practice the republican principles and Christian values that had been frustrated and defeated in his earlier life. All but household goods have been nationalized and every citizen is guaranteed comfortable and equal support throughout his life. This material security has precipitated a revolution in social assumptions and morals. The vices of egotism have vanished and a new altruism, based upon recognition of the organic nature of society, has sprung up in their place. The ideal of public virtue and the spirit of social cohesion have at last become a reality. West is introduced into the new society by Dr. Leete, his wife, and daughter Edith, who discovered his sleeping chamber and revived him from his trance. His narrative thus becomes the story of his conversion from nineteenth-century skepticism to belief in twentieth-century republicanism. Though he feels conscious of his cultural inferiority, as an anachronistic remnant of a barbarous age, the Leetes welcome him into their household as an equal. His challenge is mental and spiritual: to adjust to his new condition and to reach beyond the confines of self to the fraternity of society. Toward the end of the book his belief is tested when in a nightmare he returns to his nineteenth-century life. Revisiting scenes he once readily accepted, he is now tormented by his new perception of society's fierce competition, widespread suffering, and callous egotism. He bursts in upon a dinner party at his fiancée's home and pleads with the guests to heed the cries of the poor and build a new egalitarian social order. Unmoved, they denounce him as a mad fanatic and start to throw him out of the house. However, West reawakens in the year 2000, but now happily engaged to Edith Leete, who has revealed that she is the great-granddaughter of his lost love Edith Bartlett. Their romance provides him with a link to his past life and a surrogate for his loss and at the same time signals his integration into the new society.

In a series of conversations which form the bulk of *Looking Backward*, Dr. Leete explains to West how society achieved the republican ideal. As the industrial monopolies which West had known swelled to gigantic proportions by the early twentieth century, a new "national party" sprung up, advocating government ownership and operation of these trusts for the good of the whole people. Its members contended that nationalization of industry would raise both society's productive capacity and its ethical level as citizens worked in a vital union for the common good. This revitalized republicanism gradually won more and more converts as private monopolies absorbed one another. The process finally ended in the nationalization of industry and capital as the people of the United States extended the principles of the American Revolution from politics to industry. As Dr. Leete presents it, socialism marks the culmination of republicanism.

The economic basis and school of republicanism in Bellamy's Nationalist utopia is the "industrial army," in whose ranks both men and women must serve between the ages of twenty-one and forty-five. Here governmental and working forces are combined in a minutely structured hierarchy, beginning with common laborers and rising to the general of the army, the President of the United States. As a youth Bellamy himself had dreamed of a military career, an ambition that was dashed when he failed the physical examination for entrance to West Point. Nonetheless, he retained a lifelong fascination with military affairs, amusing himself even in his final illness in 1898 by deploying toy soldiers on his bed.[31] For Bellamy the military offered a model of both social ethics and organization, of solidarity and efficiency. With William James, Oliver Wendell Holmes, Jr., and other members of their generation, he keenly admired the martial virtues of selflessness, discipline, valor, patriotism and sought to find, in James's phrase, "a moral equivalent of war" to train a healthy citizenry. In Bellamy's industrial army, workers learn "habits of obedience, subordination, and devotion to duty" and are encouraged toward their best efforts by appeals to their sense of honor and service.[32] As the industrial army instructs workers in republican virtue, it also forges them into a mighty economic machine. A centrally coordinated system of planning, production, and distribution completely supersedes the wasteful and erratic economy of the nineteenth century to achieve unprecedented abundance. For Bellamy the war on poverty is hardly a metaphor. At one point Dr. Leete argues that the comparative effectiveness of the divided organization of manpower under private capital as against the unified industrial army may be likened to "the military efficiency of a mob, or a horde of barbarians with a thousand petty chiefs, as compared with that of a disciplined army under one general—such a fighting machine, for example, as the German army in the time of Von Moltke." Evidently Julian West is persuaded by Leete's analogy, for in his nightmare return to nineteenth-century Boston the only admirable sight he finds is the appearance of a military parade: "Here at last were order and reason, an exhibition of what intelligent cooperation can accomplish." In language that recalls Dr. Leete's, he sees the passing regiment not as individuals but as a "tremendous engine...able to vanquish a mob ten times as numerous," and he marvels that the spectators do not see in its scientific organization and centralized control a model for the rest of society.[33] Thus for Bellamy as for a number of other utopian and dystopian writers, the ultimate extension of technological organization lay in the mechanization of men themselves.

To help mold workers into this industrial machine, Bellamy devised a system of industrial management that anticipated in rudimentary form some of the features of Frederick W. Taylor's system of scientific management. (Indeed Taylor would have made an excellent general in Bellamy's industrial army.)[34] Promotions through the ranks of the industrial army are intensely competitive, and, instead of increased wages, proper motivation is encouraged through elaborate manipulations of status, including privileges, prizes, and honorable mentions. Dr. Leete rather smugly observes that the "nobler sort of men" rise above this spirit of emulation and obey their own sense of duty—though this opinion is suspect, since he later confesses his own youthful ambition to win an honorific ribbon. These positive incentives are reinforced by punitive pressures. A special presidential police, called the inspectorate, supervises all aspects of the industrial army. "Not only is it on the alert to catch and sift every rumor of a fault in the service, but it is its business, by systematic and constant oversight and inspection of every branch of the army, to find out what is going wrong before anybody else does."[35] Those workers who persist in negligence or disobedience are placed in solitary confinement on bread and water until they relent. The criminally deviant are regarded as exhibiting "atavistic" behavior and their illness treated in hospitals.

Further to ensure discipline, members of the industrial army are not allowed to vote for their commander-in-chief, the President. Only after they have been mustered out of the ranks at age forty-five do they gain the franchise that permits them to indicate a preference for President from among the ten lieutenant-generals who command various departments. Thus in Bellamy's utopia, society as a whole has become a kind of republican asylum. As in earlier republican asylums such as Lowell, factory discipline and adherence to the industrial hierarchy form the basis of republican deportment. Since the possibility of conflicts of interest is denied, no provision is made for political dissent. Indeed, Bellamy sought to eliminate politics altogether and to reduce the complex structure of government and industry alike to machine-like simplicity, efficiency, and regularity. Although municipal governments still exist in his utopia, state governments have become superfluous. Congress meets only every five years, rarely considers significant legislation, and in any case may only commend new bills to the following session "lest anything be done hastily." Instead of governmental innovation the system requires only superintendence; instead of politics, management. Administrators of the industrial army, Dr. Leete assures West, need have no more than fair abilities: "The machine which they direct is indeed a vast one, but so logical in its principles and direct and simple in its workings, that it all but runs itself; and nobody but a fool could derange it."[36] By insisting that the economy and government of his utopia would be largely self-regulating, requiring only technical adjustments rather than decisions of value and acts of authority, Bellamy hoped to mitigate the irony that to achieve the ends he most desired in a technological republic, he sacrificed, or at least severely circumscribed, democracy itself.

Bellamy no doubt would argue that this curtailment of political liberty is not really a sacrifice at all, since no substantial differences are presumed to exist, and in return his utopia promises complete social and economic security, even unprecedented abundance. In fact, Bellamy built this exchange of libertarian surrender and material reward into the heart of his social system. Beyond the privileges and honors to be gained from advancement within the industrial army lie the even greater joys of retirement. Once workers have discharged their duties and are mustered out of the industrial army at age forty-five, unless they assume one of the high administrative posts they are free to spend what Dr. Leete calls "the brighter half of life" at leisure. As to *how* they spend it, Leete gestures grandly to the cultivation of "the higher exercise of our faculties, the intellectual and spiritual enjoyments and pursuits

Mark Twain's "mechanical miracle," the Paige Compositor.

which alone mean life."[37] Somewhat anticlimactically, however, he acknowledges that the majority of retirees do not share these artistic, scientific, or scholarly interests. It would be surprising if they did, since they have had no real opportunity to acquire such tastes and skills; the intellectual and professional departments are run separately from the industrial army proper. Instead, most retired workers are left to participate in what appears to be the principal recreation in Bellamy's utopia: material consumption.

Images of consumption and of the technological achievements that make it possible are in fact one of the most powerfully and concretely rendered aspects of this utopia, far more so than the new sense of human solidarity and brotherhood that Bellamy insists upon throughout but never demonstrates in specifically human terms. (The reserved benignity of the Leete family is hardly an exception.) Significantly, when West first awakens in the year 2000 and refuses to believe that he really has slept 113 years, Dr. Leete does not attempt to convince him by demonstrations of the new social ethic but by the physical spectacle of the new Boston, a model City Beautiful:

> Miles of broad streets, shaded by trees and lined with fine buildings, for the most part not in continuous blocks but set in larger or smaller inclosures, stretched in every direction. Every quarter contained large open squares filled with trees, among which statues glistened and fountains flashed in the afternoon sun. Public buildings of a colossal size and an architectural grandeur unparalleled in my day raised their stately piles on every side.

Faced with this material grandeur, West is immediately convinced of the reality of his great transition: "Surely I had never seen this city nor one comparable to it before."[38] Bellamy's city represents "the metropolis as a department store"; it embodies an almost exclusively consumer culture.[39] Though Bellamy gestures vaguely to the existence of clubhouses for vacations and sports, he shows no public social interaction. He concentrates instead upon depicting the city's two great institutions of consumption: the local dining-hall and the distribution center (where finished goods are purchased by credit card). Both offer efficient, impersonal service in sumptuous surroundings. Over the entrance to the latter stands "a majestic life-size group of statuary, the central figure of which was a female ideal of Plenty, with her cornucopia," the symbol of Bellamy's promise of abundance for all and the key to his society's cohesion.[40] The problem of luxury, Bellamy suggested, disrupted American society in the nineteenth century only because it reflected gross inequities. Equally shared and guaranteed abundance eliminates the desire to hoard possessions.

This culture of consumption has transformed domestic life in general. Though families still maintain individual residences, all such tasks as washing, cooking, and sewing are performed in public facilities. But as domestic chores have been appropriated by the state, so have domestic pleasures, and a technological mass culture placed in their stead. To take a prophetic example, the singing and playing of music in the home have been totally superseded by professional concerts transmitted by a kind of telephonic radio. Church services, similarly, are rarely attended in person but rather

audited and sampled by this medium. In his fascination with technological devices as stimuli, Bellamy even anticipates the clock
radio and Muzak, with programmed reveilles in the morning.[41] In
general, responsibility, interaction, and hence emotional intensity
among members of the family have been parceled out to the state.
Though the husks of domesticity remain, both public and private
passions have been diffused to a benign, brotherly glow.

The modern reader's response to Bellamy's utopian vision is
inevitably tinctured by the fact he too is "looking backward" and
that as more and more of the specific features of Bellamy's society
have been realized, the ideal grows less attractive. As Lewis Mumford has observed, utopias such as Bellamy's, with their standardiza-
tion, isolation, stratification, and militarism, merge all too readily
into the dystopias, both fictional and actual, of the twentieth century.[42] It is easy, if perhaps unfair, to see Bellamy's utopia as
culminating in fascism and the Orwellian tyranny of *1984* or
dovetailing insidiously into Huxley's *Brave New World*. But the
fears that Bellamy's book raises are not prompted solely by
twentieth-century experience. Almost all the reservations modern
critics have expressed concerning the militarism and materialism of
Bellamy's utopia were previously voiced by his own contemporaries.[43] What made Bellamy's work controversial both in his own
time and in the twentieth century was the way in which he expressed the tensions and dilemmas of the quest for an ideal technological republic. Castigating both the material and ethical results
of industrial capitalism, Bellamy attempted to reassert popular control over technology in order to achieve a society where republican
virtue once again could flourish. In this spirit he boldly projected a
socialist order consonant with American traditions and experience.
Significantly, he insisted upon the word "nationalist" rather than
"socialist," a word he "never could well stomach." Bellamy explained his aversion to the term in a letter to William Dean
Howells: "In the first place it is a foreign word in itself and equally
foreign in all its suggestions. It smells to the average American of
petroleum, suggests the red flag, with all manner of sexual novel-
ties, and an abusive tone about God and religion, which in this
country we at least treat with decent respect."[44] Bellamy wished
not to set one class against another but to enlist all citizens of good

will in a campaign to abolish class itself; he aimed not to foment a new revolution but to defend the ideals of the Revolution of 1776 by extending popular control from government to industry. Beneath this egalitarian strain, however, his conception of republicanism still contained strong elements of the hierarchical structure, institutional discipline, and social control which had been important components in conservative republican thought ever since the late eighteenth century. Seizing power from a capitalist elite, he thrust it into the hands of a technocratic one. His faith in the unlimited potentialities of human nature and of technology once they were freed from the prison of nineteenth-century capitalism led him to assume that when these bars were lifted mankind could return to its original innocence and build a heaven on earth. He did not see that his own system could rigidify into a prison itself.

"Began 'Looking Backward' Nov. 5, 1889, on the train. A fascinating book," Mark Twain wrote in his notebook shortly before the publication of his own novel *A Connecticut Yankee in King Arthur's Court*. Within a month he was hailing Bellamy as "the man who has made heaven paltry by inventing a better one on earth," and soon after arranged to have Bellamy visit him at his home in Hartford, Connecticut.[45] One wonders if these authors discussed their two books, for the works make an intriguing comparison.[46] Whereas *Looking Backward* presented Bellamy's vision of an American technological republic in the future, in *A Connecticut Yankee* Mark Twain sent a contemporary American back in time to institute a technological republic in the distant past. Both novels fictionally tested the implications of a republican order based on technology, and in both cases dystopian elements arose out of their visions despite themselves. Though Mark Twain was less of a conscious critic of nineteenth-century industrialism than Bellamy, *A Connecticut Yankee* presents some of the most disturbing portents of the failure of the utopian quest for a technological republic of any book in American literature.

Study of Mark Twain and *A Connecticut Yankee* is especially illuminating because he beheld technology with his culture's most uncritical fascination and yet uttered some of his era's gravest forebodings over the course it was taking. His father was a reckless speculator who reportedly dreamed of devising a perpetual motion machine, and early in life Mark Twain was fired with the spirit of technological enterprise.[47] A would-be inventor and entrepreneur, he provided during the course of his career financial support for as many as a hundred inventions and manufacturing schemes, almost all of them unsuccessful. These projects included: a steam generator, a steam pulley, a marine telegraph, a watch company, an engraving process, a carpet-pattern machine, a Telelectroscope (an early sort of television), a skimmed milk cure-all called plasmon, a cash register, and a spiral hat pin. Mark Twain even held patents of his own for three (deservedly obscure) inventions: an adjustable and detachable clothing strap, a pregummed scrapbook (from which he actually made money), and a memory-building game.[48] As a writer, he was particularly fascinated by inventions that facilitated communication. He owned the first private telephone in Hartford, bought and used one of the first Remington typewriters, avidly collected fountain pens, and experimented with phonograph dictation.[49] This enthusiasm led him from 1880 to 1894 to the most famous and disastrous of all his technological obsessions, his enormous investment of time, money, and energy in James W. Paige's typesetting machine.[50]

The Paige typesetter was one of a number of machines in the nineteenth century designed to imitate and supersede a human printer by setting, justifying, and distributing single foundry types automatically. An impossibly cumbersome and intricate device, it weighed about 5000 pounds and contained some 18,000 separate parts, 800 shaft bearings, and innumerable springs.[51] By 1884 the invention was already rendered obsolete by Ottmar Mergenthaler's Linotype machine, which instead of distributing type, melted it down and started afresh after each run. However, Mark Twain obstinately refused to acknowledge Mergenthaler's superiority, and the deeper his obsessive support of Paige took him, the wilder his visions of immense wealth. At times his pages of figures prophesied returns approaching the billion mark, and he gleefully exulted that he would need ten men to count the profits. "I am one of the wealthiest grandees in America—one of the Vanderbilt gang, in fact," he boasted in 1890, yet ruefully added he had hardly a dollar to spare.[52]

A master of popular rhetoric, Mark Twain was also its victim. The language of the technological sublime shut his ears to the absurdity of his quixotic venture. Paige, he insisted, in phrases that recall Thomas Ewbank, was "a poet; a most great and genuine poet, whose sublime creations are written in steel. He is the Shakespeare of mechanical invention." To the typesetter itself, Mark Twain paid extravagant tribute, attributing to it human skills and personality. Thus he urged his friend William Dean Howells (like himself, a onetime printer) to attend a demonstration in October 1889: "You & I have imagined that *we* knew how to set type—we shabby poor bunglers. Come & see the Master do it! Come & see this sublime magician of iron & steel work his enchantments." The anthropomorphic conception of the machine that ran through so many of his comments became explicit as he boasted in another letter that his "magnificent creature of steel" was in construction "as elaborate and complex as that machine which it ranks *next*, to, by every right—Man—and in performance it is as simple and sure."[53] The extraordinary complexity of the machine should have warned him of its impracticality; but in his technological naïveté that very complexity proved irresistibly appealing, another instance of the mysterious wonders of invention. In one of the many false dawns when Mark Twain thought the typesetter finally stood on the brink of perfection, he wrote his brother Orion Clemens in January 1889, grandly announcing that "at 12:20 this afternoon a line of movable types was spaced and justified by machinery, for the first time in the history of the world!" Drunk with the epochal significance of the machine, he continued:

> All the other wonderful inventions of the human brain sink pretty nearly into commonplace contrasted with this awful mechanical miracle. Telephones, telegraphs, locomotives, cotton gins, sewing machines, Babbage calculators, Jacquard looms, perfecting presses, Arkwright's frames—all mere toys, simplicities! The Paige Compositor marches alone and far in the lead of human inventions.[54]

During much of the period when Mark Twain was most actively engaged in the development of the Paige typesetter, from

December 1884 to May 1889, he conceived and wrote *A Connecticut Yankee*. In his enthusiasm and frustration, he even developed an irrational sense of linkage between the novel and the typesetter and wished to complete both projects on the same day, as if he too were a word-machine.[55] In a sense, the book may be regarded as an attempted justification (though ultimately a judgment) of Mark Twain's own passionate involvement with technology. By sending a nineteenth-century Yankee (in the early notebook entries the character was Mark Twain himself) to Arthurian England and depicting his adventures, he intended a comic contrast between two cultures: modern, republican, technological America and primitive, aristocratic, superstitious England. In conception, at least, the novel defended contemporary American society against both millennialist critics and nostalgic dreamers. History was a record of ethical, political, and technological progress from medieval barbarism to the glories and comforts of the present. America would advance toward perfection by developing along existing lines. Thus Mark Twain ostensibly offered his readers a reassuring message. He confirmed the achievements of the present by journeying back in time to burlesque the romantic attraction to the Middle Ages. He made this point explicit in an unpublished preface: "If any are inclined to rail at our present civilization, why—there is no hindering him, but he ought to sometimes contrast it with what went before and take comfort and hope, too."[56]

Mark Twain's protagonist, Hank Morgan, describes himself at the outset of the story as "a Yankee of the Yankees—and practical; yes, and nearly barren of sentiment, I suppose—or poetry in other words." He stands as the embodiment of nineteenth-century technological man, scornful of the traditional arts and fascinated by machinery. At the Colt arms factory in Hartford (where Mark Twain first saw the Paige typesetter), Morgan boasts he "learned all there was to it; learned to make everything; guns, revolvers, cannon, boilers, engines, all sorts of labor-saving machinery." The capacity for violence suggested by his work he cheerfully acknowledges, saying he is "full of fight."[57] Gradually he has risen from worker to manager. As head superintendent at the factory, he backs up his official authority over the workers by sheer physical strength

—until in a duel with crowbars a worker hits Morgan over the head so hard he sends him back in time thirteen centuries to Arthurian England.

Morgan's situation is thus the reverse of Julian West's in *Looking Backward*; so too is his response. The Yankee is so confident of his internal stability that at first he believes all the inhabitants of Camelot to be simply inmates in a lunatic asylum. When he discovers that he indeed has awakened in the year 528, he quickly gets over his astonishment and adjusts to his new situation. To a believer in the progressive course of history, a plunge backward to the sixth century would seem to be unendurable torture, but as a practical Yankee he discovers personal compensations: "Look at the opportunities here for a man of knowledge, brains, pluck and enterprise to sail in and grow up with the country. The grandest field that ever was; and all my own; not a competitor; not a man who wasn't a baby to me in acquirements and capacities." Far better was it from an entrepreneurial point of view, Morgan decides, than if (like West) he had been hurled forward in time to the more advanced twentieth century; for there he "could drag a seine down-street any day and catch a hundred better men than myself." In the Nationalist order of the year 2000, West discovered to his astonishment that even the gold in his vault was worthless. Morgan, by contrast, contemplates his position in the Middle Ages as an extraordinary financial opportunity, feeling "just as one does who has struck oil."[58]

Morgan thus approaches Arthurian England like a nineteenth-century American industrialist before a newly discovered country populated by a curious, technologically backward, and therefore "primitive," people. Or in his words, "I saw that I was just another Robinson Crusoe cast away on an uninhabited island, with no society but some more or less tame animals, and if I wanted to make life bearable I must do as he did—invent, contrive, create, reorganize things; set brain and hand to work, and keep them busy. Well, that was in my line."[59] A one-man imperial expedition, the Yankee regards Arthur's subjects much as nineteenth-century English and Americans viewed underdeveloped nations of black Africans and Indians of their own time. On different occasions he calls them "white Indians," "modified savages," "pigmies," "great simple-hearted creatures," "big children," "sheep." They are pictured as credulous, superstitious, gleeful, innocent, cruel, irrational, dirty, vulgar—all the "native" virtues and vices. The Yankee, by contrast, regards himself as clever, practical, rational, foresighted, chaste, and humane—the virtues of a technological and republican culture, though they are virtues which will themselves be tested in the course of the novel.

As a stranger from another culture, the Yankee's position in Arthur's realm resembles that of the shipwrecked European sailor who, before the advent of the white man, washed up on the Javanese coast, was taken by the people for a white monkey and chained to a rock. Though he scratched his name and an account of his shipwreck in three languages, he never really communicated with his captors. As they doubted his humanity, he may have indeed doubted theirs.[60] Hank Morgan is taken by the English, if not for a white monkey, at least for a kind of monster or powerful animal, certainly not a man like themselves. Condemned to be burned at the stake, the Yankee forestalls his execution with what had become in nineteenth-century literature a well-established mode of dealing with "savages," whether American Indians, black Africans, or, in this case, medieval English. Morgan providentially remembers the imminent occurrence of a total eclipse of the sun and, armed with this bit of scientific trivia, he stages a "miracle."[61] As a sharp Connecticut trader, Morgan threatens to blot out the sun unless he is made King Arthur's chief minister and promised one per cent of the revenue his programs contribute to the state. His success in achieving this agreement confirms his sense of superiority.

Once established in power, the Yankee institutes a program of technological development and political reform in order to recast sixth-century England in the image of nineteenth-century America. In this effort, as in the case of so many ventures in the nineteenth century, motives of republican reform and personal aggrandizement combine. Morgan covertly establishes a network of industries and communications throughout the land and trains a cadre of technological experts to administer them. Like the founders of Lowell and other model factory towns, he conceives of his industrial complex as a cradle of republican civilization. Reversing Victorian critics of factory life, preeminently John Ruskin, who contrasted the modern operative as "an animated tool" with the freedom of the medieval craftsman, Morgan commends his factory as a place "where I'm going to turn groping and grubbing automata into *men*."[62]

Within its comic framework, then, *A Connecticut Yankee* raises the question of the true civilizing power of Morgan's technology and his republican system. He aspires to be a Great Man who will literally change the course of history.[63] But as with all utopian schemes, there is the formidable problem of how to institute the new order. On this issue, despite his cocksure manner, Morgan remains uncertain. As he temporizes in pointless farcical quests or (like so many late nineteenth-century reporters and sociologists) joins with the king in incognito excursions to discover "how the other half lives," he debates how to initiate his program of reform. At first Morgan rejects a sudden transition in favor of "turning on my light one-candle-power at a time." Though he defends the Reign of Terror as infinitely swifter and more humane than the thousand-year terror of the *ancien régime*, still he shrinks from revolution, arguing he must educate "his materials" first. The mute resignation of the people, however, at times makes him despair of ever succeeding. Fatalistically, he declares:

> All gentle cant and philosophising to the contrary notwithstanding, no people in the world ever did achieve their freedom by goody-goody talk and moral suasion: it being immutable law that all revolutions that will succeed, must *begin* in blood, whatever may answer afterward. If history teaches anything, it teaches that. What this folk needed, then, was a Reign of Terror and a guillotine, and I was the wrong man for them.[64]

Here Morgan epitomizes his ambivalent position, denouncing mere progressive rhetoric and declaring only violent revolution will answer, then shrinking from his own political imperative—though finally he will embrace it with a vengeance. At the root of his strategic dilemma is his contradictory position as a self-styled republican reformer who feels no ideological unity with the people he proposes to lead, only condescension and contempt. Occasionally, he gives some recognition to this problem. A number of times in the course of the book he speaks forcefully about the inflexibility of established attitudes and the impossibility of over-turning the effect of cultural training overnight. Nevertheless, in the last analysis he perceives all the English people's departures from his own cultural values as evidences of primitivism. In benign moods, he regards them as "the quaintest and simplest and trust-ingest race" and boasts that he stands among them as "a giant among pigmies, a man among children, a master intelligence among intellectual moles: by all rational measurement the one and only actually great man in that whole British world." Occasionally, he discovers an individual who responds to his influence, declares him "a man"—his highest tribute—and sends him to his factory settle-ment. But at other moments he sighs despairingly, "There are times when one would like to hang the whole human race and finish the farce."[65]

Ultimately, the Yankee is a person more attracted to the idea of achieving what he calls his "new deal" than to republicanism itself. This "deal," the plan to industrialize and democratize Arthu-rian England, arises more out of his own needs and assumptions than those of the English themselves. His deepest ambition is personal: to be "the greatest man in the kingdom," to exert "enormous authority," to remake Arthurian England in his own image. He revels in his title spontaneously awarded by the people, which, "translated into modern speech, would be THE BOSS." This is not the title of a republican leader but that of a dictator, a phrase linking him to the political and industrial bosses of the nineteenth century and, still more ominously, translated into modern Italian and German, to il Duce and der Führer of the twentieth.[66] Despite his professions of republicanism, the Yankee displays alarming fondness for despotic power, as when he exults over the swift development of his factories: "Unlimited power is the ideal thing when it is in safe hands."[67]

How "safe," then, are Morgan's hands? A number of passages in the novel raise basic questions as to the character of his mission. As he vaunts his clandestine factory colony, his language contains a strong note of menace: "There it was, as sure a fact, and as sub-stantial a fact as any serene volcano, standing innocent with its smokeless summit in the blue sky and giving no sign of the rising hell in its bowels."[68] This use of volcanic imagery and mixture of praise and dread in describing the Yankee's nascent industrial civilization is continued in the name of the newspaper he establishes: the "Camelot *Weekly Hosannah and Literary Volcano*."[69] The ambiguous title points to a fundamental dichotomy in Morgan's venture. Is he a Yankee Saviour bringing technological and political enlightenment in order to lead the people to a new heaven on earth? Or is he an Exterminating Angel, prefiguring the character Satan in Mark Twain's *The Mysterious Stranger*, destined to destroy them? The mode of Morgan's public displays of technological power reinforces doubts. Whether blowing up Merlin's tower with a lightning rod and blasting powder or repairing a holy fountain with pipe, a pump, and a grand display of fireworks, he depends for his success both upon his own practical knowledge and his audience's continued ignorance. As much as his rival Merlin, he is a magician, concealing his methods behind a subterfuge of showmanship and special effects.[70] His displays are more threatening than constructive, calculated to arouse the public's terror and admiration and to boost his reputation as he battles Merlin for power. Furthermore, once he can amass the materials, these "miracles" reveal the Yankee's irresistible penchant for explosives and growing disregard for human life. When first he actually kills with his devices, using a dynamite bomb against two mounted knights, it is ostensibly to save the king's life. Nevertheless, he has been fairly itching for an excuse, and he revels in the spectacle of destruction like a boy at a fireworks: "Yes, it was a neat thing, very neat and pretty to see. It resembled a steamboat explosion on the Mississippi; and during the next fifteen minutes we stood under a steady drizzle of microscopic fragments of knights and hardware and horse-flesh." The image of nineteenth-century technological disaster raises further doubts as to the efficacy of Morgan's program. Later, the lethal capacity beneath his American folk humor asserts itself more baldly as the Yankee challenges the members of the Round Table in a self-proclaimed battle "to either destroy knight-errantry or be its victim." He begins his series of jousts with a lasso, dressed like a circus performer, and ends an arrogant and ruthless gunman exulting over ten dead.[71]

With this triumph the Yankee at last feels free to unveil his burgeoning nineteenth-century civilization and to develop it openly. Within three years, he proclaims, "Slavery was dead and gone; all men were equal before the law; taxation had been equalized. The telegraph, the telephone, the phonograph, the typewriter, the sewing machine, and all the thousand willing and handy servants of steam and electricity were working their way into favor. We had a steamboat or two on the Thames, we had steam war-ships, and the beginnings of a steam commercial marine; I was getting ready to send out an expedition to discover America."[72] Morgan plans to overthrow the Catholic Church in favor of Protestant denominationalism and after Arthur's death at last to institute his republic—with himself as first president. Yet one sees nothing of real improvements in popular life; quite the opposite, in fact: undisciplined speculation in the stock market the Yankee has established soon divides the Round Table and leads to civil war. The analogy to irresponsible and criminal financial manipulations in Mark Twain's own time and their polarizing effect upon American society is clear.[73] Ironically, the Yankee has reproduced nineteenth-century American civilization all too faithfully, and he sighs, "My dream of a republic to *be* a dream, and so remain."[74] The hope that technological civilization could lift Arthurian England out of its morass of ignorance, inequality, and oppression to republican enlightenment and virtue fails; and by implication, it signals the failure of technology to institute a true republic in America. The issuing of typewriters and sewing machines to the people does not automatically usher in the millennium. Instead, Morgan's experiment backfires: the Church exploits the anarchy his schemes have created to entrench its power more firmly than ever and to ban the Yankee and modern technology.

When Morgan finally does proclaim a republic, it is an empty gesture, calculated to provoke a fight. The people in whose name he pretends to speak reject his program, and his only remaining supporters are his assistant Clarence and fifty-two adolescent boys, all of whom have been schooled since childhood in his factories and thoroughly indoctrinated in his principles. As the Yankee's vision is at last decisively rejected, it turns to destructive megalomania. Throughout, Morgan has reflected his experience at the arms works in a strong affinity for military technology and a fascination with battle; he has even established his own West Point.[75] As he turns to gain by force what he has failed to achieve by peaceful means, his actions reveal the technological violence of which Americans (among others) are capable when their republican values are opposed by an alien and technologically less advanced people, such as the American Indians in Mark Twain's own time or the Vietnamese in recent years. Failing to win "the hearts and minds" of the medieval English by his program of industrial development or by his limited duels, the Yankee escalates the conflict to a war against all comers, against all forces of cultural resistance. The war of liberation thus becomes a war of extermination. Morgan installs his crew in a fortress and lovingly and meticulously assembles his era's most modern technology of death: Gatling guns, land mines, and, his *pièce de résistance*, a row of electrified fences.[76] These weapons consummate the interest Morgan expressed in such "labor-saving machinery" at the very outset of the story. For they were all regarded in the late nineteenth century as more efficient and hence, proponents reasoned, more "humanitarian" weapons because they would lead to smaller armies and shorter wars. The inventor of the Gatling gun, Dr. Richard J. Gatling, who lived not far from Mark Twain in Hartford and whose gun Mark Twain had delightedly test-fired at the Colt arms works as early as 1868, defended his weapon in just these terms, and so too did enthusiastic defenders of dynamite guns and land mines in popular magazines.[77] Because of this increased efficiency, technological innovations in weaponry were particularly celebrated as at last assuring the supremacy of the forces of civilization over their "barbarian" antagonists.[78] The Gatling gun, never fired in the American Civil War for which it was developed, became a favorite weapon of the United States army in the Indian wars of the 1870s and 1880s as well as of the British army in the Zulu war of 1879.[79] Morgan's actions at the end of *A Connecticut Yankee* are thus less of an aberration than critics have suggested. Having regarded the Round Table as "a sort of polished-up court of Comanches" and the population in general as "white Indians," he proceeds to treat them as such.[80]

A Connecticut Yankee ends in an Armageddon between the forces of the nineteenth century and the sixth which becomes a study in technological atrocity. Morgan defines it as a war to end wars, fought in the name of the republic, a crusade on behalf of liberty and equality, a struggle between animal might and the resources of free and intelligent men. But in fact it is a war between two clashing systems of oppression. If the English forces marching against Morgan represent the barbarism of feudalism and a corrupt Church, the Yankee himself comes to represent the authoritarianism and dehumanization of uncontrolled technological power. Although he is ostensibly fighting on behalf of technological civilization, he blows up his own beloved factories lest they fall into enemy hands. The final legacy of his inventiveness is not his "civilization-factories" but an automated battlefield. His sophisticated weaponry insulates him and his youthful assistants from direct contact with the enemy and as a consequence he loses all sense of restraint.[81] He delights in his power to kill efficiently and distantly through his technology and indulges in a feeble military wit, what might be called the pornography of destructive power, which conflates extraordinary and deadly actions with mundane and innocent ones. Thus Morgan and his assistant Clarence speak of the "music" of Gatling guns and offer mock hospitality toward the enemy. In one exchange, the Yankee savors the news of how a party of clerics who marched toward the fortress to demand their surrender has "tested" the land mines:

> "Did the committee make a report?"
> "Yes, they made one. You could have heard it a mile."
> "Unanimous?"
> "That was the nature of it."[82]

In his fascination with technological destruction, Morgan, who has earlier been appalled by the torture of a single man, discovers an awful joy in his ability to kill eleven thousand with a single switch of an electric fence: "*There* was a groan you could *hear*!" His scorn of the English as a low and undifferentiated "mass" at last achieves its physical correlative. "Of course we could not *count* the dead," he explains at one point after exploding a land mine, "because they did not exist as individuals, but merely as homogeneous protoplasm, with alloys of iron and buttons."[83] Nevertheless, he does announce a body count of 25,000—a final application of the quantitative standards of technological "progress." As the perpetrator of technological atrocity, the Yankee is not only physically insulated but emotionally desensitized. He counters this "psychic numbing" to a degree only when he leaves his fortress to confront the dead and wounded directly. Once on the battlefield, he is exposed physically and also emotionally. The wound he receives from an injured enemy is the product and symbol of his residual guilt.[84]

The book concludes heavy with irony. Morgan's youthful supporters are trapped in their fortress and die infected by the rotting bodies they have killed. Clarence pronounces the moral of their technological assault: "We had conquered; in turn we were conquered."[85] Merlin's magic ultimately triumphs over the Yankee's as he enters the fortress in disguise and places Morgan under a spell to sleep thirteen centuries. At last, then, after an age-long sleep, the Yankee returns to his once beloved nineteenth century. Significantly, however, unlike Julian West, he is not reintegrated to society in the end. Instead, he dies in delirium, longing for his Arthurian wife as an image of restoration, a symbol of an imagined harmonious past which he feels he has betrayed.[86] His dream of introducing a republic by fomenting an industrial and political revolution has turned into a nightmare, and not simply loss but a deep sense of unresolved guilt are implicit in his final words to his now centuries-dead wife: "Death is nothing, let it come, but not with those dreams, not with the torture of those hideous dreams—I cannot endure *that* again."[87]

The full depths of Mark Twain's fable of technological and political progress were ignored by American reviewers of *A Connecticut Yankee*, who generally took the Yankee's self-declared position as a humanitarian and republican industrialist at face value. Bellamy's supporter Sylvester Baxter enthusiastically applauded the book as "eloquent with a true American love of freedom, a sympathy with the rights of the common people, and an indignant hatred of oppression of the poor, the lowly and the weak." He even bemoaned the fact that savages similar to the medieval English still existed among the Indians of the American West, and in central Asia and Africa, who had not been exposed to "the light of a genuine civilization."[88] Mark Twain's friend William Dean Howells heralded the novel in *Harper's Monthly* as "an object-lesson in democracy" which "makes us glad of our republic and our epoch."

As late as 1908 Howells "re-re-read" *A Connecticut Yankee* and wrote Mark Twain, "It is the most delightful, truest, most humane, sweetest fancy that ever was."[89]

Though a number of recent critics have argued that Mark Twain's difficulties in control of the book reveal his internal frustrations and ambivalent attitude toward the ideology of technological progress, it is difficult to determine to what extent he acknowledged the import of his novel.[90] In any case, to Mark Twain more than to most writers applies D. H. Lawrence's dictum: "Never trust the artist. Trust the tale."[91] While Edward Bellamy's *Looking Backward* castigated the economic and social cleavages he believed inherent in industrial capitalism, Mark Twain's *A Connecticut Yankee* challenged his society's progressive values on another, still more disturbing level. The book demonstrated how a powerful, supposedly humanitarian republican leader may betray his own ideals as he seeks to extend control over a weaker, underdeveloped nation through essentially aggressive use of his technology. This tendency, present throughout the book, explodes into genocidal violence at the end. The inability of reviewers to confront the theme of technological atrocity at the heart of the novel only reveals how deep-seated the Yankee's capacity for violence and for the self-deception which supported it was shared by society at large.

The author of *Caesar's Column*, Ignatius Donnelly, surged with a passion for social justice and an apocalyptic imagination congenial to Mark Twain's own. But while Mark Twain grew increasingly pessimistic of the prospect of social reform, Donnelly aimed to enlist his readers in a massive campaign of social regeneration. Known variously as "The Sage of Nininger," "Prince of Cranks," "Apostle of Discontent," and "Tribune of the People," Donnelly achieved a colorful and multi-faceted career as a reformer and rebel. He participated actively in politics for over forty years and further pursued his interests and assaults on orthodoxy as a lawyer, farmer, lecturer, lobbyist, political journalist, scientific popularizer, Baconian, and social novelist. His greatest political success was his first. Moving to Minnesota from his native Philadelphia in 1856, he joined the newly emergent Republican party and was elected Lieutenant Governor by the age of twenty-eight. He served two terms in the state house, followed by three more in Congress as a Radical Republican; but in 1868 he was squeezed out by the Minnesota Republican organization and defeated in his try for a fourth term. Donnelly won election to the state legislature on a number of occasions in following years. However, he never regained his Congressional seat, let alone the Senate post he coveted. Instead, he assumed a leading role in the many third-party movements that swept the Midwest in the late nineteenth century—Grangers, Greenbackers, Farmers' Alliance, Populists—and gradually emerged as one of the region's most important agrarian spokesmen. In 1900, despite failing health, he ran for Vice-President on the Mid-Road People's party ticket, those Populists who refused merger with the Democrats. Thus a rebel to the end, he died on January 1, 1901.[92]

Donnelly voiced the protest of an agrarian America suddenly eclipsed by the industrial transformation of the late nineteenth century. Instead of easing the plight of farmers and laborers, the rapid proliferation of technology appeared to him and other Populists only to increase their burden. By the early 1890s they joined in demanding radical changes in the capitalist structure and popular control of technology, particularly through nationalization of the railroads.[93] The conviction that America stood once again in the balance between aristocratic subversion and a new assertion of republican principles seized them with apocalyptic intensity. Donnelly vividly expressed this sense of social crisis in his famous preamble to the Populist platform of 1892. "We stand," he declared, "in the midst of a nation brought to the verge of moral, political and material ruin. . . . The fruits of the toil of millions are boldly stolen to build up colossal fortunes, unprecedented in the history of the world, while their possessors despise the republic and endanger liberty. . . . A vast conspiracy against mankind has been organized on two continents and is taking possession of the world. If not met and overthrown at once it forbodes terrible social convulsions, the destruction of civilization, or the establishment of an absolute despotism." Echoing the language of the Declaration of Independence and the Preamble to the Constitution as radical workers had done ever since the 1830s, Donnelly called for a new revolution in American politics that would restore the republic to "the 'plain people' with whom it originated."[94]

Although Donnelly was a master of political oratory, even such fervid and grandiloquent rhetoric did not do justice to the urgency of the crisis he foresaw and the vividness of his imagination. To dramatize fully the alternatives facing America, he did more than attend reform conventions; he turned to the genre of utopian fiction which Bellamy's *Looking Backward* had popularized. Here, freed from the exigencies of the political platform, he revealed most starkly his views on the nightmarish and utopian alternatives of American political and technological development. Donnelly wrote *Caesar's Column* quickly in 1889 in the wake of another defeat in his quest for the United States Senate. At least five different publishers rejected it for its "revolutionary not to say inflammatory character," and one of them pleaded with Donnelly that if he insisted on publishing the book at least to price it above a dollar lest it fall into "the wrong hands." Donnelly did not heed this advice, however. *Caesar's Column* finally appeared in 1890 under a pseudonym in both hardcover and paperback editions. The novel attracted over 60,000 American buyers during its first year of sales; by 1899 their numbers had risen, according to Donnelly's estimate, to 230,000, plus another 450,000 in Europe. The national constituency that Donnelly was continually denied as an office-holder, he attained as a social novelist.[95]

Donnelly's novel offers a double vision of utopia and catastrophe in ways that link it to both Bellamy's *Looking Backward* and Mark Twain's *A Connecticut Yankee*. Though in his conclusion Donnelly presented his own version of utopia, the negative vision, the dystopia, dominates the book. In many respects he played off the expectations of Bellamy's readers, allowing them to reach his paradise only after a purgative journey into hell.[96] The book depicts the adventures of Gabriel Weltstein, a shepherd of Swiss extraction from Uganda, visiting New York in 1988. At first he is as dazzled by the city as Julian West is by the new Boston. Glass-covered arcades, elegant shops, elevated and underground electric railroads, and ingenious airships all compel his admiration.

The dining room of his sumptuous hotel surpasses even the luxurious facilities West encountered. In a triumph of mechanical service, guests summon exotic foods to their table at the press of a button and eat in silence, entertained by a global telecommunications system.

However, as Donnelly releases the springs of his plot, Weltstein quickly discovers that this society is the opposite of utopia. Even in his hotel he senses that material power has flourished at the expense of individual worth and human charity. Soon after, he finds himself innocently embroiled in an international revolutionary movement called the Brotherhood of Destruction. From one of the revolutionary leaders, Maximilian Petion, he learns that America remains a republic in name only. Despite the trappings of representation, it is in fact ruled by a tyrannous merchant, Prince Cabano, and the Council of the Oligarchy. The palace from which the prince exerts his malevolent authority is the embodiment of luxury which Americans ever since the Revolution had been taught to fear. Filled with rare books, precious statuary, kingly paintings, and other works of art, it represents "the very profligacy and abandon of unbounded wealth."[97] As the crowning mark of his debauchery, the prince keeps a bevy of mistresses. His latest acquisition to his harem, the innocent and uncomprehending Estella Washington, a direct descendant of "the father of our country," symbolizes the plight of republican virtue menaced by the ravishment of tyranny.

Both this private luxury and the public splendor which the city affords the rich depend upon the ruthless oppression of the majority. Petion takes Weltstein on a tour of the houses and factories of the working class, in effect a separate city called the "Under-World." There he discovers a stunted and enslaved proletariat, reduced to a deathlike submission, so that they have almost ceased to be human and have become "automata." Donnelly's depiction of their brutalized and numbed existence anticipates later descriptions of the Nazi concentration camps. Even their end is chillingly prophetic: they are swept off in scores as soon as dead, carted away in streetcars, and thrown into white-hot furnaces, finally to vanish through a high chimney. "That towering structure was the sole memorial monument of millions of them. Their graveyard was the air."[98]

The barbarism West encountered in *Looking Backward* was that of his own nineteenth century before the principles of Nationalism have triumphed. Weltstein, by contrast, confronts a society that has continued to deteriorate past all possibility of social reconstruction. Cataclysm can no longer be averted. "Two hundred years ago," Petion tells Weltstein, "a little wise statesmanship might have averted the evils from which the world now suffers. One hundred years ago a gigantic effort, of all the good men of the world, might have saved society. Now the fire pours through every door, and window and crevice; the roof crackles; the walls totter; the heat of hell rages within the edifice; it is doomed; there is no power on earth that can save it." Though farmers and laborers can still muster the strength to revolt against their rulers, they are too degraded to create a new civilization. The Christian social message that knits the citizens of Bellamy's utopia in a common bond has here all but ceased to exist. The rich have embraced a religion of cruel oppression and egotistical hedonism, and the poor have lost all hope in the possibility of redemption. "We have ceased to be men—we are machines," a labor spokesman exclaims to Weltstein. "Did God die for a machine? Certainly not."[99]

Politically bankrupt, socially polarized, morally decayed, the nation rushes toward catastrophe. Prince Cabano and his councillors coolly plot to crush the festering rebellion and retrench their power by killing ten million people. The Brotherhood of Destruction, meanwhile, moves to foil the prince's plot and massacre the aristocracy. When the revolutionaries manage to bribe the air force and thus gain control of their dynamite and poison bombs, they have the crucial advantage. As in *A Connecticut Yankee*, then, the culmination of this society's technological development is to facilitate its own destruction. By having the idealistic Weltstein accompany the revolutionist Petion through the final stages to the revolution itself, Donnelly allows the reader to seek vicarious revenge over the corrupt and despotic plutocracy even while standing morally opposed and aghast. The division in Mark Twain's Hank Morgan between humanitarian ideals and revolutionary bloodthirstiness is thus in Donnelly's story objectified (if diluted) in two separate characters. Weltstein watches in fascinated horror as the elite air corps, called the "Demons," use their advanced technology to destroy the prince's foot soldiers. As the oppressors at last stand defenseless before the mob, the revolution turns to anarchic looting and killing. Like the Angel Gabriel, for whom he is named, Weltstein presents the slaughter of the rich as a last judgment for their greed and complacency:

> You were blind, you were callous, you were indifferent to the sorrows of your kind. The cry of the poor did not touch you, and every pitiful appeal wrung from human souls, every groan and sob and shriek of men and women, and the little starving children—starving in body and starving in brain—rose up and gathered like a great cloud around the throne of God; and now, at last, in the fullness of time, it has burst and comes down upon your wretched heads, a storm of thunderbolts and blood.[100]

The decline of the American republic from agrarian virtue to plutocratic imperialism to mob barbarism is encapsulated in the life-history of the leader of the revolutionary Brotherhood, Caesar Lomellini. His very name, indeed, recalls the earlier decline and fall of Rome. Once he had been "a quiet, peaceful, industrious" farmer living happily in the newly settled state of Jefferson. However, he fell in debt to bankers, who enslaved him and his family, drove them from their home, and seduced and abandoned his daughter. The treachery of the plutocrats transforms Caesar from a peaceful farmer into a self-appointed avenger of the poor and enslaved. He launches a rebellion against the economic oppressors that culminates in the international revolution of the Brotherhood of Destruction. As Prince Cabano personifies the spirit of plutocratic tyranny, so Caesar embodies the beastlike violence of the unleashed mob.[101] Together they represent the two opposing threats to republican society, aristocratic tyranny and popular demagoguery, which Americans ever since the Revolution had been taught to fear. In the midst of the slaughter of the rich, Caesar boasts that now *he* is king and drunkenly urges on the killing. Finally, he is confronted with the same problem that Hank Morgan faces in *A Connecticut Yankee*, of how to dispose of the corpses before his own forces perish through infection. Despite his drunkenness, he displays an inventiveness the Yankee would admire. Caesar orders his men to erect a cement column of the dead, laced with dynamite to discourage tampering. Weltstein pronounces the column a monument to "the death and burial of modern civilization."[102] As the anarchic passions of the mob sweep out of control, Caesar ultimately falls victim of the revolution he has helped to create. He last appears in the novel with his head on a stake. The collapse of American civilization is complete, and the golden age of the early republic will rise again, if at all, only in a millennium.

However, Donnelly does not conclude on this cataclysmic note. Though the United States and Europe have succumbed to barbarism, Weltstein and his friends manage to escape in a last airship and flee back to Uganda. America no longer offers the pastoral setting requisite for a healthy republic, and Africa emerges as the last "new world." Nevertheless, this retreat does not signify a final rejection of modern technology and all its works in favor of the shepherd's pipes. Before Weltstein departs, he assembles a kind of Noah's ark of the most important elements of Western civilization: "literature, science, art, encyclopedias, histories, philosophies, in fact all the treasures of the world's genius—together with type, printing presses, telescopes, phonographs, photographic instruments, electrical apparatus, eclesions, phemasticons . . ."—Donnelly adds the names of fictitious inventions to encompass twentieth-century developments. Once they have returned safely to their "garden in the mountains," Weltstein quickly sets about instituting a new republican order to protect his homeland from the destructive fate of republics abroad. His first step is to barricade the entrances to his country from the outside world and to manufacture rifles and cannon to repel invaders. Then, Weltstein and a few other "superior intellects" devise a constitution which the people accept. It is in some ways an elitist document, dividing the governing body into three branches by profession: workmen and farmers; merchants, manufacturers, and employees; and writers, artists, scientists, and philosophers. In this republican utopia technology plays an important but carefully circumscribed role. Weltstein's experience has made him deeply suspicious of uncontrolled development of manufactures and industry. Instead, he hopes to divert attention from the development of isolated inventions to more humanly significant social problems. "If the same intelligence which has been bestowed on perfecting the steam-engine had been directed to a consideration of the correlations of man to man, and pursuit to pursuit, supply and demand would have precisely matched each other, and there need have been no pauperism in the world."[103] Toward this end the state itself assumes control over technology. It owns all railroads, mines, and utilities. Though it does not actively encourage new labor-saving inventions, it buys rights to useful ones for the benefit of the people. In all cases, its principle is to subordinate technology to the welfare of the republic.

Thus, Donnelly and the Populist thought he reflects did not oppose technology as such. On the contrary, he affirmed that freed from the imperatives of a capitalist and competitive order and guided by an enlightened socialism, modern machinery might play a significant role in the liberation and fulfillment of mankind. Nevertheless, a certain sense of uneasiness pervades Donnelly's inclusion of technology within his utopia, a concern to keep it within carefully prescribed limits. In this respect his pastoral society contrasts markedly with Bellamy's technocratic order. Whereas Bellamy seized upon the capacity of modern technology to create a society of unprecedented abundance, Donnelly clung to an older conception of a harmonious and egalitarian society arising out of rusticity and an economy of sufficiency rather than abundance. These were the essential preconditions of a pastoral republic as he understood it and the reason why technology was necessarily subordinate. His vision of a "garden in the mountains," free from external contamination and composed of village workmen and yeoman farmers powerfully recalls the pastoral ideal of *Notes on Virginia*. Yet writing a century after Jefferson, Donnelly could not heedlessly indulge in the Virginian's enthusiasm for labor-saving devices without any sense of their potential threat to his pastoral ideal. Instead, Donnelly attempted to achieve the social order which Jefferson thought would naturally develop through strict governmental regulation of technological development, industrialization, and urbanization. While Jefferson relied upon the ecology of liberty to support his republican garden, Donnelly found it necessary to build a hothouse and to place within it a *memento mori*, as a continuing reminder of the catastrophic fate of uncontrolled technological republics.

The crisis of American republicanism in an urban-industrial age and the possibility of utopian reconciliation, considered in such feverish and apocalyptic tones by Ignatius Donnelly, were also explored in the early 1890s by a very different kind of social novelist, William Dean Howells. The two writers make a striking juxtaposition. While Donnelly characteristically assumed a grandiloquent, even quixotic, attitude, Howells impressed his contemporaries with his dignified, reasonable, unassuming appearance. While Donnelly functioned both in politics and in literature as a rebel on the periphery, Howells by the early 1890s stood squarely at the center of the world of literary affairs, "the dean," in the popular tribute punning on his name, "of American letters." Despite their contrasting personalities and social positions, however, Howells and Donnelly converged in their concern that an industrial plutocracy was fast replacing democracy in America and frustrating the expression of republican values. In turning to utopian fiction, both writers attempted to break the chain between republican ideology and the imperatives of industrial capitalism that had been continually forged and strengthened since the early nineteenth century and to recover a vision of agrarian and egalitarian America that the nation had lost.

As recent scholarship has amply shown, Howells was a far more complex and serious writer than the figure H. L. Mencken contemptuously dismissed as "a placid conformist" who tepidly produced "a long row of uninspired and hollow books, with no more ideas in them than so many volumes of the *Ladies' Home Journal*."[104] Though he rose with dizzying swiftness from his Ohio boyhood to the summit of the American literary establishment, Howells always retained an essential ambivalence toward this literary world and his own critical and financial success in it.[105] His agreeable and prosperous bearing masked a troubled sensibility, increasingly given to doubting both his own achievement and the course of American society as a whole. As he confided to Henry James in 1888, "I'm not in a very good humor with 'America' myself. . . . I should hardly like to trust pen and ink with all the audacity of my social ideas; but after fifty years of optimistic content with 'civilization' and its ability to come out all right in the end, I now abhor it, and feel that it is coming out all wrong in the end, unless it bases itself anew on a real equality. Meantime, I wear a fur-lined overcoat, and live in all the luxury my money can buy."[106] This passage, with its barely controlled frustration and self-lacerating irony, explodes the stereotypical image of Howells as an exemplar of gentility and complacency. Moreover, he was not content to fume in private; he *did* trust his pen with his social ideas both in his novels in the late 1880s and 1890s and in regular columns for *Harper's Monthly* and other popular magazines.[107] More than any other American novelist of comparable stature at the end of the nineteenth century, he used his voice and influence in the cause of social reform, from publicly protesting the impending execution of the anarchists convicted of throwing a bomb in Chicago's Haymarket Square to commending in magazine editorials a wide variety of important social critics ranging from Tolstoy to Veblen. In addition, Howells also began to feel the lure of utopia, the need of a social model with which to exert sufficient leverage to lift American society back to its republican ideals. He expressed interest in a number of the cooperative experiments of the day, including W. D. P. Bliss's Christian Socialist mission in Boston and Edward Bellamy's Nationalist party. Nevertheless, he shrunk from discipleship in any organized crusade, keenly aware of his ambiguous position as a prosperous citizen attempting to bridge class divisions by preaching equality and understanding. Fascinated as he was by the example of Tolstoy, who renounced his aristocratic privileges for the life of a peasant, Howells found it impossible to emulate. As he ruefully remarked of himself, Mark Twain, and their wives, "We are theoretical socialists, and practical aristocrats."[108] This gulf between practice and conviction, Howells came to realize, resulted not simply from personal hypocrisy but from the false positions into which all Americans were forced by the existing social and economic order. The difficulty of escaping the horns of this social dilemma and of successfully acting upon reformist convictions formed the theme of some of his most important novels of this period, including *Annie Kilburn* (1888) and *A Hazard of New Fortunes* (1890). In an effort to clarify this theme and to dramatize the gulf in American society as a whole between social ideals and social practice, Howells, like so many other writers of this period, turned to utopian fiction in the early 1890s and wrote *A Traveler from Altruria*.

In tone and setting, though not in theme, *A Traveler from Altruria* differs significantly from *Looking Backward*, *A Connecticut Yankee*, and *Caesar's Column*. Instead of placing his characters in strange surroundings in the past or the future, Howells locates them squarely in the present in the comfortably familiar setting of a New Hampshire summer hotel. There they are visited by a traveler from the distant country of Altruria, Mr. Homos (whose name, Greek for "same," suggests equality and homogeneity). The Altrurian presents himself as a curious though uninitiated student of American civilization, familiar with the Declaration of Independence and other expressions of American republican ideals and eager to study their workings firsthand. His host at the hotel is a writer of genteel romances, Mr. Twelvemough. Twelvemough quickly emerges as Howells's portrait of the artist as a moral coward. He characteristically views life in terms of literary effects rather than human involvement and attempts to mask with jokes and quips his basic timidity and insecurity. Twelvemough introduces Mr. Homos into the circle of business and professional men staying at the hotel, who, like himself, are defined in terms of their class and occupations. In conversations with one another they begin by complacently repeating the democratic pieties of equal opportunity and political equality. However, under Mr. Homos's persistent probing, they gradually acknowledge that despite having "a republican form of government, and manhood-suffrage, and so on," America is permeated with economic and social inequality. They jealously protect their economic self-interest by suppressing their generous emotions as inappropriate to the world of affairs and exercising brute power against the working class in order to smash the unions and thwart their demands. Their operative principle in business has nothing to do with equality or justice, certainly nothing to do with charity; it is, "The good of

Number One first, last and all the time."[109] Behind the façade of a democratic society, their conversations reveal, exists a ruthless scramble for economic and social dominance.

As Howells presents it, these men are engaged in a race without winners, only losers. Despite their social privileges and affluence, they are almost as trapped and driven in their lives and work as the poor. "You can have no conception of how hard our business men and our professional men work," boasts Mr. Twelvemough, thinking all their frenetic activity admirable. "There are frightful wrecks of men strewn all along the course of our prosperity, wrecks of mind and body. Our insane asylums are full of madmen who have broken under the tremendous strain, and every country in Europe abounds in our dyspeptics." While these men are exhausted by the pressures of work, their wives are similarly driven to nervous collapse by the dictates of conspicuous leisure and consumption. Mrs. Makely, the voice of the leisure class in the novel, catalogues the endless round of a society lady's duties: "making herself agreeable and her house attractive, ... going to lunches, and teas, and dinners, and concerts, and theatres, and art exhibitions, and charity meetings, and receptions, and ... writing a thousand and one notes about them, and accepting and declining, and giving lunches and dinners, and making calls and receiving them." She finally shrieks to the Altrurian, "It's the most hideous slavery! You don't have a moment to yourself; your life isn't your own!"[110]

In such moments members of the upper class profess to envy the simple life of the workers, and this sentimental solicitude is the nearest the two classes come to true communication. For their part, the working class is hedged in by economic tyranny on every side. They farm the poor land until it slides to the total control of the banks, work in the mills to be laid off during depressions, or join the desperate migration to the cities and the West. Without condoning their circumstances, Howells takes the hopeful view that the poor, at least, far more than the wealthy, approach a condition of true equality and brotherhood, if only for the simple reason that "poor people have always had to live [for one another], or they could not have lived at all."[111]

Throughout these conversations, Mr. Homos has given Americans only glimpses of Altrurian society, saying he wishes to learn rather than teach. Finally, however, he agrees to give a benefit lecture on Altruria to both the prosperous guests of the hotel and the farmers and factory workers of the surrounding area. His talk quickly emerges as a parable of America's historic development and a prophetic vision of the way to utopia. Once (like America) Altruria had overthrown a monarch and instituted a republic. Anticipating a new era of freedom and brotherhood, citizens discovered new energies of mind and hand. New inventions sprang forth and commerce flourished, promising prosperity and the public good. Yet the era of industrialization proved to be a new tyranny:

> It was long before we came to realize that in the depths of our steamships were those who fed the fires with their lives, and that our mines from which we dug our wealth were the graves of those who had died to the free light and air, without finding the rest of death. We did not see that the machines for saving labor were monsters that devoured women and children, and wasted men at the bidding of the power which no man must touch.[112]

The Accumulation, as this power was called, grew and consolidated its strength until it owned almost the entire means of production. But as in Bellamy's *Looking Backward*, in this monolithic condition it proved most vulnerable to nationalization when citizens finally realized its true monstrousness. In a peaceful and legal revolution, the Accumulation was transformed into the commonwealth of Altruria.

Altruria represents Howells's conception of the true fulfillment of republicanism, the flowering of Christian community, and the integration of the machine. In a number of respects it resembles Bellamy's utopia: citizens live for one another in conditions of perfect equality; there is no competition for goods, no money, no politics, no war, no crime. Yet one of Howells's principal objections to the utopia of *Looking Backward* was that it conceded too much to materialist taste. "I should have preferred," he later wrote of Bellamy's book, "if I had been chooser, to have the millennium much simpler, much more independent of modern inventions, modern conveniences, modern facilities. It seemed to me that in an ideal condition (the only condition finally worth having) we should get along without most of these things, which are but sorry patches on the rags of our outworn civilization, or only toys to amuse our greed and vacancy."[113] Bellamy wished to free technology from the inefficiencies of capitalism. Howells, by contrast, wished to break free of the dominance of technocratic values altogether—of expansion, quantity production, specialization, efficiency, regimentation, centralization—and to reintegrate the machine into an organic way of life. Or as Mr. Homos expresses it:

> Our life is so simple and our needs are so few that the handwork of the primitive toilers could easily supply our wants; but machinery works so much more thoroughly and beautifully, that we have in great measure retained it. Only, the machines that were once the workmen's enemies and masters are now their friends and servants; and if any man chooses to work alone with his own hands, the state will buy what he makes at the same price that it sells the wares made collectively. This secures every right of individuality.[114]

Like Ruskin in *Unto This Last*, Howells in effect asserted, *"There is no Wealth but Life."* "Labor-saving" inventions formerly served not actually to save labor but to increase production, profits, and consumption, resulting in extremes of wealth and poverty and a materialistic society in which people had luxuries but not necessities. In Altruria, however, a balanced economy ensures sufficiency of consumption for all, and the superfluous products of an endlessly expanding economy are forbidden. Since men no longer compete with one another economically, Mr. Homos repeatedly emphasizes, *"there is no hurry."* Altrurians work only three hours a day. Whatever efficiency of production may be lost is compensated by a greater creative life. Freed from artificial pressures, men and women rediscover "the pleasure of doing a thing beautifully." Art and industry at last are reconciled; but this marriage is achieved not by celebrating brute power, material conquest, and wealth in the spirit of the technological sublime, or by elevating the inventor at the artist's expense. People are not distinguished by their occupations in Altruria. All work daily in factory or field, and in the less intense pace of work, all may be artists: "In all Altruria there was not a furrow driven or a swarth mown, not a hammer struck on house or on ship, not a stitch sewn or a stone laid, not a line written or a sheet printed, not a temple raised or an engine built, but it was done with an eye to beauty as well as to use."[115]

The revitalized character of work in Altruria is part of a general assimilation of modern technology within a new pastoralism. Instead of the complexity and intense centralization of *Looking Backward*, Howells celebrated a simplified, decentralized society in which man is integrated with his physical and social environment. Altrurians have abandoned their old cities altogether, preserving them only as a historic and admonitory lesson of the chaos and corruption of a former age. People now live either in regional capitals or villages, and farming is again as in Jefferson's Virginia honored as the occupation closest to God and most conducive to that love of home, family, and community which stands at the center of Altrurian values. Fewer desires or needs impel men to travel from one place to another; and although swift electric transportation operates in the capitals and along major routes, most of the old railroads that shackled the landscape have been deserted and recovered by nature. Altruria's small communities encourage free and spontaneous social life, including festivals, picnics, plays, and sports; yet each individual may have as much privacy as he desires. In short, individual character, variety, happiness, and disappointment still exist, but in no case are they determined by economic conditions. In his vision of a humane technological republic, freed from the dominance of capitalistic values, Howells looks both back to Emerson and a host of nineteenth-century critics and forward to twentieth-century formulations of an organic society based upon modern technology, of which Lewis Mumford is perhaps the most famous proponent.[116]

However, Howells intentionally presented only a sketch of Altruria, not a blueprint. His parable offered a way of sidestepping certain issues while raising others. Like most utopian writers, he only vaguely described how the Altrurian society arose out of the grim regime of the Accumulation. Similarly, his description of life in Altruria has a generalized, spectatorial air. He gestures to factories and shops as "temples" of art and industry but does not actually show us a worker before a machine. Howells concentrated upon the ethical dimension of his utopia rather than the functional. When he attempted to flesh out his depiction of Altrurian society in a sequel, *Through the Eye of the Needle* (1907), he lost the power of the Altrurian parable without appreciably adding to his utopia's solidity. However, Howells anticipated a mixture of hope and incredulity from his readers. They stand in the position of the hotel guests at the end of *A Traveler from Altruria* as Mr. Homos leaves to visit New England's farms and factory towns, wondering if he is not an impostor, if Altruria is not too good to be true. "I feel as if he were no more definite or tangible than a bad conscience," Mr. Twelvemough remarks at one point; and that, of course, is precisely the ambition Howells had for Altruria as a whole: to act as the nagging conscience of America.[117] He insisted that as a physical goal it would always remain elusive. Howells took pains to qualify his utopia because he was well aware of the fatal arrogance characteristic of most utopian thought. He wished Altruria to serve not as a precise pattern for reform but as an animating possibility and informing ideal.

Looking Backward, A Connecticut Yankee, Caesar's Column, and *A Traveler from Altruria* mark the culmination of over a century of American attempts to integrate technology and republican values. As Bellamy, Mark Twain, Donnelly, and Howells presented their visions of technological utopia and dystopia, they encapsulated many of the issues Americans had addressed ever since the Revolutionary era: the possibility of forging an alliance between technology and republican ideology that would achieve the vision of the Revolution; the problem of establishing factories that would serve as republican communities; the impact of technology upon imaginative freedom; and the desire to fuse art and machine technology as the basis of a republican aesthetic. No more than previous generations did these writers achieve a final synthesis in the dialectic between technology and republicanism. Howells's remark to Charles Eliot Norton after completing his sequel to *A Traveler from Altruria* applies to them all: "I have given my own dream of Utopia, which I fancy your not liking, unless for its confessions of imperfections even in Utopia. All other dreamers of such dreams have had nothing but pleasure in them; I have had touches of nightmare."[118] In fact, the way in which nightmare intrudes upon the dreams of a utopian order forms a recurrent theme among the utopian novelists of this period. Rightly seen, Bellamy suggested through the experience of his character Julian West, the existing conditions of nineteenth-century American society were nightmare enough; but critics found still more disturbing the technocratic authority of his utopia. In his book *Caesar's Column*, Donnelly was consumed with depicting the cataclysmic destruction of America and the industrialized world as a whole if his countrymen allowed the fleeting opportunity to control the nation's unbridled industrial development and economic polarization finally to slip from their grasp. Most disturbing of all is the way in which dream turns horribly into nightmare in *A Connecticut Yankee*. The idea of a comic excursion of a nineteenth-century American into Arthurian England first came to Mark Twain quite literally in a dream.[119] But as Hank Morgan assumes his role of emissary of technological republicanism to medieval monarchy, the narrative exposes the brutality, dehumanization, oppression, and self-deception latent in his culture. The centuries-long nightmare in which Morgan dies haunted by guilt and the need for forgiveness signaled in a way readers were unable to acknowledge the confusion of technological and republican values in nineteenth-century America as a whole.

Despite the nightmarish possibilities of American technological civilization which concerned all these writers and which Mark Twain expressed most deeply, the dream of a humane technological order persisted. The utopias of Bellamy, Donnelly, and Howells, for all their important differences, reflect a common effort: to revitalize the egalitarian strain of American republicanism as an ideological basis for a new social order. They contended that the economics and ethics of industrial capitalism, while mutually reinforcing, were neither desirable nor inevitable. To locate an alternative tradition and source of value, they insisted, one need not turn to the disruptive radicalism preached in Europe; the values of the Revolutionary generation and their heirs, of Jefferson and Emerson, while obscured and thwarted in the uncontrolled expansion of urban-industrial society, still could offer Americans inspiration and guidance as they approached a new millennium. The Revolutionary vision of a new era of human freedom, political and moral purity, social responsibility and harmony, these novelists affirmed, far from being a dead dream, remained a vital possibility quite within their grasp. The spirit of commercialism, competition, and individualism which stifled cooperative effort and altruistic impulses arose not out of brute "human nature," but from an inequitable and exploitative economic system. Bellamy, Donnelly, and Howells all strenuously advocated extending the principle of democratic control from government to industry as the essential step in achieving a truly republican order. If ever competition for material goods was necessary, it was no longer so in an age of modern technology. Government control of the means of production could provide abundance for all. Technology itself had outgrown the competitive capitalist ethic which clung to it and demanded a new cooperative republican socialism. Only allow technology to serve the people as a whole rather than benefit a few special interests and peace, abundance, leisure, equality, and virtue could be assured. Thus while the key instrument of Bellamy, Donnelly, and Howells for attaining utopia was government control of technology, the heart of their concern was moral. They demanded for each citizen an equitable share of society's goods not as the ultimate fruit but as the essential precondition for utopia. Once established on a common ground with a genuine community of interests, former plutocrats and proletarians would at last be reconciled, and the generous, creative, and noble aspects of humanity would flower as never before. While Donnelly, the only politician among these writers, allowed his profession a limited role even in utopia, Bellamy and Howells planned their republics to be so united as to banish politics altogether, as a relic of the vulgarity and corruption of the late nineteenth century.

Yet the very purity for which these writers ached, the way in which they wished, in one great leap, to transcend the divisive economic, social, and political conditions of the late nineteenth century, was a measure of the obstacles before them. For most Americans of all classes the vital sense of connection between republican ideology and American society was no longer clear in the polarized industrial order of the late nineteenth century. Earlier efforts to unite the two had lost their guiding force. Lowell's directors had long ago abandoned their experiment of a model republican community. Emerson was dead, his radicalism tamed, his doctrine of self-reliance perverted to the dogma of laissez-faire. Rather than pursuing his critical evaluation of technology's impact upon republican civilization, middle- and upper-class Americans were increasingly content to view the course of American technological development as a grand, progressive pageant, a national work of art. Like the hotel guests in Howells's *Traveler from Altruria*, they continued to pay tribute to republican ideals while supporting a social system that effectively denied their promise. As they sought to mask the contradiction between belief and action, they reduced republicanism's egalitarian thrust to complacent homilies of self-help, the notion of the public good to facile assertions of management and labor's "community of interests." The result was to convert republicanism from an animating ideology to a static buttress of the conservative industrial order.

The utopian impulse, then, was an expression of republican ideology *in extremis*, a desperate attempt to demonstrate the continuing relevance and transforming power of America's republican tradition for the crises of industrial society. Utopian novelists intended their books to serve as a diagnosis and prescription for the gap between republican ideology and industrial society. Ultimately, however, they stood as symptoms of this very split. As the republican values for which these writers stood were attenuated and thwarted in existing society, they were compelled to invent in fiction a social order in which such ideals could flourish. The utopian novel provided an arena to give substance to their ideas, to demonstrate their implications for human experience, which they were denied in the larger society. In part, too, novelistic embodiment of their visions provided an emotional compensation for their practical defeat. Bellamy identified a tendency toward wishful dreaming not only in his own character but in the entire quest for republicanism in the late nineteenth century when he wrote, "Having fully calculated upon and expected a thing, I am so justly disappointed by its failure to come to pass that the balance of my nature goes over to the potential world, and I go to Might-have-been-land."[120] In the authorial freedom of "Might-have-been-land," as Howells had recognized, writers tended to fit society to the Procrustean bed of their theories—just as influential spokesmen for the industrial order had stretched and trimmed republicanism to *their* own purposes. As a consequence of this division between dreamers of "the potential world" and defenders of the existing social order, both sides developed reductive visions, and social debate was impoverished. The creative tension between republican ideology and industrial society threatened to lapse into simple contradiction.

Thus, utopian writers could express their sense of the unrealized possibilities of technology and republicanism, but they lacked any real leverage to implement their beliefs. While they attracted large audiences, as long as the cleavage between conviction and action remained, their ideas might gain assent without exerting meaningful influence. Their views could be accepted merely as expressions of benign intentions and their books consumed as an ethical tonic of republican spirits, sweetened with Christian molasses and laced with a reproving sulphur. Between the novelists' cosmic vision and the instrumentalities they proposed to attain it still yawned an immense gulf, for all their insistence in the feasibility of the passage. Though their message would continue to exert an important influence among progressive reformers in the decades to come, never again would the concept of republicanism possess the centrality and coherence that made it, from the Revolution through the nineteenth century, such an important shaping ideology in Americans' response to technology. Instead, new political strategies and categories of value would have to be formulated to address the enduring and elusive problem of civilizing the machine.

Notes

1. The Emergence of Republican Technology

1. On ideology, see Clifford Geertz, "Ideology as a Cultural System," in *Ideology and Discontent*, ed. David E. Apter (Glencoe, Ill., 1964), pp. 47–76. On republicanism in Revolutionary America, see especially: Bernard Bailyn, *The Ideological Origins of the American Revolution* (Cambridge, Mass., 1967); Gordon S. Wood, *The Creation of the American Republic, 1776–1787* (Chapel Hill, 1969); and Robert E. Shalhope, "Toward a Republican Synthesis: The Emergence of an Understanding of Republicanism in American Historiography," *William and Mary Quarterly*, 3d ser., 29 (January 1972), 49–80.

2. Bailyn, *Ideological Origins*, pp. 55–56; Wood, *Creation of the American Republic*, pp. 46–70.

3. Bailyn, *Ideological Origins*, pp. 94–143.

4. Bailyn, *Ideological Origins*, pp. 160–319. See also Howard Mumford Jones, *O Strange New World: American Culture: The Formative Years* (New York, 1964), pp. 327–50.

5. Wood, *Creation of the American Republic*, pp. 393–429. See also John R. Howe, Jr., "Republican Thought and the Political Violence of the 1790s," *American Quarterly*, 19 (Summer 1967), 147–65.

6. Richard Hofstadter, *America at 1750: A Social Portrait* (New York, 1971), pp. xi, 3–32, 131–79.

7. *De Officiis*, I, xii (Loeb trans.), quoted in A. Whitney Griswold, *Farming and Democracy* (New York, 1948), pp. 19–20. See also Paul H. Johnstone, "In Praise of Husbandry," *Agricultural History*, 11 (April 1937), 80–95; Bailyn, *Ideological Origins*, pp. 23–26.

8. J. Hector St. John de Crèvecoeur, *Letters from an American Farmer* (1782; rpt. London, 1908), pp. 78–79. On agrarian nationalism in eighteenth-century America, see three articles by Chester E. Eisinger: "The Freehold Concept in Eighteenth-Century American Letters," *William and Mary Quarterly*, 3d ser., 4 (January 1947), 42–59; "The Influence of Natural Rights and Physiocratic Doctrines on American Agrarian Thought During the Revolutionary Period," *Agricultural History*, 21 (January 1947), 13–23, and "Land and Loyalty: Literary Expressions of Agrarian Nationalism in

the Seventeenth and Eighteenth Centuries," *American Literature*, 21 (May 1949), 160–78.

9. See *Oxford English Dictionary*, s.v. "technology."

10. *The Papers of Benjamin Franklin*, ed. Leonard W. Labaree *et al.* (New Haven, 1959–), II, 380–83; Brooke Hindle, *The Pursuit of Science in Revolutionary America, 1735–1789* (Chapel Hill, 1956), pp. 68–69; I. Bernard Cohen, "How Practical Was Benjamin Franklin's Science?" *Pennsylvania Magazine of History and Biography*, 69 (1945), 284–93.

11. Edmund S. Morgan, "The Puritan Ethic and the American Revolution," *William and Mary Quarterly*, 3d ser., 24 (January 1967), 3–43.

12. Victor S. Clark, *History of Manufactures in the United States*, 1929 ed. (New York, 1929), I, 31–72, 188–90; Rolla Milton Tryon, *Household Manufactures in the United States, 1640–1860* (Boston, 1891), II, 731–735; William B. Weeden, *Economic and Social History of New England, 1620–1789* (1917; Augustus M. Kelley rpt. New York, 1966), pp. 53–55. The fullest account of the boycott movements and their impact upon domestic manufactures is Arthur Meier Schlesinger, *The Colonial Merchants and the American Revolution, 1763–1776* (New York, 1918).

13. "A Speech delivered in Carpenter's Hall, March 16th...," *Pennsylvania Evening Post*, I (April 11 and 13, 1775). On the United Company of Philadelphia, see William R. Bagnall, *The Textile Industries of the United States* (Cambridge, Mass., 1893), I, 63–72; Clark, *History of Manufactures*, I, 190–91.

14. Worthington Chauncey Ford, ed., *Journals of the Continental Congress, 1744–1789* (Washington, D.C., 1906), IV, 224.

15. Clark, *History of Manufactures*, I, 219–27; Curtis P. Nettels, *The Emergence of a National Economy, 1775–1815* (New York, 1962), pp. 40–44; Don Higginbotham, *The War of American Independence: Military Attitudes, Policies, and Practice, 1763–1789* (New York, 1971), pp. 303–309.

16. I. Bernard Cohen, "Science and the Revolution," *Technology Review*, 47 (April 1945), 374. Cohen observes that Washington's experience was instrumental in his desire to establish a United States Military Academy for the training of military engineers, "our first national scientific institution."

17. Brooke Hindle, *David Rittenhouse* (Princeton, 1964), pp. 125–30, 164; Charles Coleman Sellers, *Charles Willson Peale* (New York, 1969), pp. 114–18.

18. *The Complete Writings of Thomas Paine*, ed. Philip S. Foner (New York, 1945), II, 1135–36, 1045; *The Literary Diary of Ezra Stiles*, ed. Franklin Bowditch Dexter (New York, 1901), I, 497–98; Hindle, *Pursuit of Science*, pp. 244–45; Higginbotham, *War of American Independence*, pp. 310–12; F. W. Lipscomb, *Historic Submarines* (New York, 1970), pp. 14–16; Robert Fulton, *Torpedo War and Submarine Explosions* (1810; facsimile rpt. Chicago, 1971), p. 57.

19. Wood, *Creation of the American Republic*, pp. 393–429; Morgan, "Puritan Ethic," pp. 35–38; Boston *Independent Chronicle*, August 24, 1786, quoted in Wood, p. 418; St. George Tucker, *Reflections on the Policy and Necessity of Encouraging the Commerce of the Citizens of the United States of America ...* (Richmond, 1785), pp. 7–8.

20. *Notes on the State of Virginia*, ed. William Peden (1787; rpt. Chapel Hill, 1955), pp. 164–65.

21. Richard Wells, *A Few Political Reflections submitted to the Consideration of the British Colonies* (Philadelphia, 1774), quoted in Jones, *O Strange New World*, p. 204; Adams to John Luzac, September 15, 1780, in Adams, *Works*, ed. Charles Francis Adams, VII (Boston, 1852), 255; Benjamin Franklin, "Comfort for America, or remarks on her real situation, interests, and policy," *American Museum*, 1 (January 1787), 6.

22. Hugh Williamson, *American Museum*, 2 (August 1787), 114, 118; "On Americans Manufactures," *American Museum*, 1 (February 1787), 124; "An Oration delivered at Petersburgh...," *American Museum*, 2 (November 1787), 420–21.

23. Clark, *History of Manufactures*, I, 61–62, 40–46.

24. Williamson, *American Museum*, 2 (August 1787), 118, 129; *American Museum*, 2 (August 1787), 165–67; *American Museum*, 3 (January 1788), 89.

25. Eisinger, "Influence of Natural Rights and Physiocratic Doctrines," pp. 20–22; George Logan, *Letters Addressed to the Yeomanry of the United States* (Philadelphia, 1791), p. 15, quoted in Eisinger, "Influence of Natural Rights and Physiocratic Doctrines," p. 21; Franklin, *Writings*, ed. Albert Henry Smyth (New York, 1907), V, 101–102, 202. On Jefferson and American pastoralism, see Leo Marx, *The Machine in the Garden: Technology and the Pastoral Ideal in America* (New York, 1964), pp. 116–144. The phrase "Nature's nation" is Perry Miller's; see his essay "Nature and the National Ego," collected in *Errand into the Wilderness* (Cambridge, Mass., 1956), pp. 204–16, and also (as "The Romantic Dilemma in American Nationalism and the Concept of Nature") in *Nature's Nation* (Cambridge, Mass., 1967).

26. "On American Manufactures," *American Museum*, 1 (February 1787), 122, 124. On the economy of nature, see Daniel J. Boorstin, *The Lost World of Thomas Jefferson*, Beacon Press (Boston, 1960), pp. 41–53.

27. I draw this summary from David S. Landes, *The Unbound Prometheus: Technological Change and Industrial Development in Western Europe from 1750 to the Present* (Cambridge, 1969), pp. 80–87, 41–42.

28. Landes, *Unbound Prometheus*, pp. 97–103, 65*n*; James Boswell, *Life of Johnson*, quoted in Marx, *Machine in the Garden*, p. 145. See for example Jacob Bigelow's retelling of the anecdote, in which he heightens the drama by having Boulton make the remark directly to George III, in Bigelow's *Elements of Technology* (Boston, 1829), pp. 5–6.

29. Jefferson to Charles Thomson, April 22, 1786, in *The Papers of Thomas Jefferson*, ed. Julian P. Boyd (Princeton, 1954), IX, 400–401. See also Jefferson's letter to Thomson of December 17, 1786, in which he relates a discussion he later had with Matthew Boulton on the advantages of steam engines. *Papers*, X, 608–10.

30. See esp. Jefferson's letter to Benjamin Austin, January 9, 1816, *Writings*, ed. Albert Ellery Bergh (Washington, 1907), XIV, 389–92.

31. Merrill D. Peterson, *Thomas Jefferson and the New Nation: A Biography* (New York, 1970), pp. 535–36; Dumas Malone, *Jefferson and the Ordeal of Liberty* (Boston, 1962), pp. 217–20; *Thomas Jefferson's Farm*

Book, ed. Edwin Morris Betts (Princeton, 1953), pp. 426–53; James A. Beard, Jr., "Mr. Jefferson's Nails," *Magazine of Albermarle County History*, 16 (1958), 47–52.

32. Jefferson to Thaddeus Kosciusko, June 28, 1812, in *Writings*, ed. Bergh, XIII, 170–71; Peterson, *Thomas Jefferson*, pp. 937–41; *Farm Book*, pp. 341–411, 464–95.

33. Quotation from Jefferson, letter to James Maury, June 16, 1815, *Writings*, ed. Bergh, XIV, 318.

34. Carroll W. Pursell, Jr., "Thomas Digges and William Pearce: An Example of the Transit of American Technology," *William and Mary Quarterly*, 3d ser., 21 (October 1964), 552; Bagnall, *Textile Industries*, pp. 84–165; George S. White, *Memoir of Samuel Slater . . .*, 2d ed. (Philadelphia, 1836), pp. 47–112.

35. Elisha Colt to John Chester, August 20, 1791, in *Industrial and Commercial Correspondence of Alexander Hamilton*, ed. Arthur Harrison Cole (Chicago, 1928), p. 10.

36. Washington to the Marquis de Lafayette, January 29, 1789, in *The Writings of George Washington*, ed. Jared Sparks (Boston, 1835), IX, 464; *Industrial and Commercial Correspondence of Alexander Hamilton*, pp. 259, 192; David Humphreys, *A Poem on Industry* (Philadelphia, 1794), p. 14; Linda K. Kerber, *Federalists in Dissent: Imagery and Ideology in Jeffersonian America* (Ithaca, 1970), p. 188n. On Hamilton's industrial town, see also Joseph Stancliffe Davis, "The 'S.U.M.': The First New Jersey Business Corporation," in *Essays in the Earlier History of American Corporations* (Cambridge, Mass., 1917), I, 349–522.

37. The significance of Evans's invention, however, went unappreciated by his contemporaries, including Thomas Jefferson. Siegfried Giedion, *Mechanization Takes Command: A Contribution to Anonymous History* (New York, 1948), pp. 38–39, 79–86; Greville Bathe and Dorothy Bathe, *Oliver Evans: A Chronicle of Early American Engineering* (Philadelphia, 1935), pp. 10–14.

38. Tench Coxe, "An Enquiry into the Principles on Which a Commercial System for the United States of America should be founded; to which are added some Political Observations connected with the Subject," *American Museum*, 1 (June 1787), 496–514. On Coxe's career, see Harold Hutcheson, *Tench Coxe: A Study in American Economic Development* (Baltimore, 1938); Bagnall, *Textile Industries*, I, 73–77. Leo Marx presents a somewhat different analysis of Coxe's thought in *The Machine in the Garden*, pp. 150–69.

39. Tench Coxe, "An Address to an assembly of the friends of American manufactures . . . ," *American Museum*, 2 (September 1787), 249–50. On the question of labor scarcity and mechanization, see H. J. Habakkuk, *American and British Technology in the Nineteenth Century: The Search for Labour-Saving Inventions* (Cambridge, 1962), esp. pp. 95–97.

40. Coxe, "An Address to . . . ," pp. 251, 253–55.

41. See Wood, *Creation of the American Republic*, pp. 471–518.

42. Benjamin Rush, "Of the Mode of Education Proper in a Republic," *Essays Literary, Moral and Philosophical*, 2d ed. (Philadelphia, 1806), pp. 7–8, 14–15. Extracts from the essay were originally published in *Gentle-*

man's Magazine, 56 (1786), part 2, 775-79; see *Letters of Benjamin Rush*, ed. L. H. Butterfield (Princeton, 1951), I, 387*n*.

43. James Madison, Federalist No. 49, and John Jay, Federalist No. 3, in *The Federalist*, ed. Benjamin Fletcher Wright, John Harvard Library (Cambridge, Mass., 1961), pp. 351, 98; Washington to John Jay, August 1, 1786, in *Writings of George Washington*, IX, 187.

44. Morgan, "Puritan Ethic," pp. 41-42; Stuart Bruchey, *The Roots of American Economic Growth, 1607-1861: An Essay in Social Causation* (New York, 1965), pp. 96-97.

45. Merrill Jensen, *The New Nation: A History of the United States During the Confederation, 1781-1789* (New York, 1958), pp. 148-53. On Jefferson's interest in ballooning, see Edwin T. Martin, *Thomas Jefferson: Scientist*, Collier Books (New York, 1961), pp. 63-67.

46. James Winthrop, *attrib.*, "The Letters of 'Agrippa,'" IV, in *The Antifederalists*, ed. Cecelia M. Kenyon (Indianapolis, 1966), p. 134; see also pp. xxix-xl; Wood, *Creation of the American Republic*, pp. 499-500.

47. Federalist No. 14, *The Federalist*, p. 153; Wood, *Creation of the American Republic*, pp. 501-506.

48. Adams to Jefferson, December 21, 1819, *The Adams-Jefferson Letters*, ed. Lester J. Cappon (Chapel Hill, 1959), II, 551. On Adams's fear of luxury see also Wendell D. Garrett, "John Adams and the Limited Role of the Fine Arts," *Winterthur Portfolio*, 1 (1964), 243-55.

49. Lyman Beecher, "The Gospel the Only Security for Eminent and Abiding National Prosperity," *National Preacher*, 3 (March 1829), 147. For Beecher's views on native manufactures, see "A Sermon delivered at Litchfield on the day of the Anniversary thanksgiving, December 2, 1819," in *Addresses of the Philadelphia Society for the Promotion of National Industry*, ed. Mathew Carey, 5th ed. (Philadelphia, 1820), pp. 261-86.

50. Max Weber, *The Protestant Ethic and the Spirit of Capitalism*, trans. Talcott Parsons (New York, 1930); Edmund S. Morgan, review of *Religion and Economic Action*, by Kurt Samuelsson, *William and Mary Quarterly*, 3d ser., 20 (January 1963), 135-40; Morgan, "Puritan Ethic"; John Cotton, quoted in Perry Miller, *The New England Mind: From Colony to Province* (Cambridge, Mass., 1953), p. 41; Wood, *Creation of the American Republic*, pp. 418-19.

51. Miller, *New England Mind: From Colony to Province*, pp. 27-52.

52. Adams to Jefferson, June 28, 1813, *Adams-Jefferson Letters*, II, 340; Jefferson to Adams, October 28, 1813, *Letters*, II, 391. See also Jefferson to Adams, September 12, 1821, *Letters*, II, 575. On the transition from a cyclical theory of history to the idea of progress, see Stow Persons, "The Cyclical Theory of History in Eighteenth-Century America," *American Quarterly*, 6 (Summer 1954), 147-63.

53. Franklin to Benjamin Vaughan, July 26, 1784, *Writings*, ed. Smyth, IX, 243; Mathew Carey, "On the Effects of Prosperity and Adversity," *Miscellaneous Essays* (Philadelphia, 1830), p. 451.

54. On this subject, see Fred Somkin, *Unquiet Eagle: Memory and Desire in the Idea of American Freedom, 1815-1860* (Ithaca, 1967), pp. 11-54.

55. Marvin Meyers, *The Jacksonian Persuasion: Politics and Belief*, Vintage Books (New York, 1960), pp. 28, 139, 12.

56. See for example Jackson's Third Annual Message, December 6, 1831, *Messages of Gen. Andrew Jackson* (Concord, N.H., 1837), pp. 127–28.

57. H. W. Bellows, *The Moral Significance of the Crystal Palace* (New York, 1853), p. 16.

58. Jones, *O Strange New World*, p. 231; see Metropolitan Museum of Art, New York City, *Emblems of Unity and Freedom*, with introduction by Holger Cahill (New York, n.d.).

59. *The Writings and Speeches of Daniel Webster*, National ed. (Boston, 1903), V, 66. Cf. David Potter's argument that the revolutionary message America has embodied for the world sprang not from its democratic ideology as much as its technological abundance. Potter repeats Franklin D. Roosevelt's reputed assertion that, if he could place one American book in the hands of every Russian, the volume he would choose would be a Sears, Roebuck catalogue. David M. Potter, *People of Plenty: Economic Abundance and the American Character*, Phoenix Books (Chicago, 1958), pp. 135, 80.

60. See for example *Eighty Years Progress of the United States* (New York, 1865), p. v.

61. Fredrika Bremer, *The Homes of the New World*, trans. Mary Howitt (New York, 1853), I, 540; see the thirteen-year-old Winslow Homer's sketch of a rocket ship, drawn in 1849 (Museum of Fine Arts, Boston), reproduced in K. G. Pontus Hultén, *The Machine As Seen at the End of the Machine Age* (New York, 1968), pp. 32–33; *The Inventor*, 1 (September 1855), 20, cited in Hugo A. Meier, "The Technological Concept in American Social History—1750–1860" (Ph.D. diss., University of Wisconsin, 1950), p. 324.

62. J. A. Etzler, *The Paradise within the Reach of All Men, without Labor, by Powers of Nature and Machinery* (Pittsburgh, 1833), pp. 1, 19. Among those Germans whom Etzler influenced was John A. Roebling, the designer of Brooklyn Bridge; see Alan Trachtenberg, *Brooklyn Bridge: Fact and Symbol* (New York, 1965), pp. 46–48.

63. On the *Lehrjahre* of Everett and his American contemporaries at Göttingen, see Orie William Long, *Literary Pioneers: Early American Explorers of European Culture* (Cambridge, Mass., 1935).

64. *Dictionary of American Biography*, s.v. "Everett, Edward"; Richard Henry Dana, Jr., *An Address Upon the Life and Services of Edward Everett* (Cambridge, Mass., 1865), p. 60.

65. On oratory as an expressive ideal, see F. O. Matthiessen, *American Renaissance: Art and Expression in the Age of Emerson and Whitman* (New York, 1941), pp. 14–24, 549–58.

66. George Ticknor, as quoted in Irving H. Bartlett, "Daniel Webster as a Symbolic Hero," *New England Quarterly*, 45 (December 1972), 497.

67. Ralph Waldo Emerson, *Works*, ed. Edward Waldo Emerson (Centenary ed., Boston, 1903–1904), X, 330–31; Perry Miller, *The Life of the Mind in America: From the Revolution to the Civil War* (New York, 1965), p. 279.

68. Edward Everett, *Orations and Speeches on Various Occasions*, 6th ed. (Boston, 1860), I, 270, 264, 260, 275.

69. Bigelow, *Elements of Technology*, pp. 6, 5.

70. Alonzo Potter, *The Principles of Science* (Boston, 1840), p. 9.

71. Everett, *Orations*, II, 71.

72. Everett, *Orations*, II, 246–47, 248.

73. Everett, *Orations*, II, 297–98.

74. On covert reservations to technological change, see Bernard Bowron, Leo Marx, and Arnold Rose, "Literature and Covert Culture," *American Quarterly*, 9 (Winter 1957), 377–86.

75. Daniel J. Boorstin, *The Americans: The National Experience*, Vintage Books (New York, 1965), pp. 101, 103–104. See also Marvin Fisher, *Workshops in the Wilderness: The European Response to American Industrialization, 1830–1860* (New York, 1967), pp. 157–58.

76. Nathaniel Hawthorne, "The Celestial Railroad," *United States Magazine, and Democratic Review*, 12 (May 1843), 515–23.

77. G. Ferris Cronkhite has observed similarities between the technological features of the celestial railroad and the Eastern Railroad from Salem, Massachusetts, to Boston, including causeways, a tunnel, and final passage into Boston by ferryboat. "The Transcendental Railroad," *New England Quarterly*, 24 (September 1951), 322. On Hawthorne and technology generally, see Henry G. Fairbanks, "Hawthorne and the Machine Age," *American Literature*, 28 (May 1956), 155–63.

78. For a stimulating discussion of the reciprocal relationships between technology and culture, see George H. Daniels, "The Big Questions in the History of American Technology," and the commentary of John G. Burke and Edwin Layton, *Technology and Culture*, 11 (January 1970), 1–35.

2. *The Factory as Republican Community: Lowell, Massachusetts*

1. Asa Briggs, *Victorian Cities*, American ed. (New York, 1965), pp. 85–135; Alexis de Tocqueville, *Journeys to England and Ireland*, trans. George Lawrence and K. P. Mayer (New Haven, 1958), pp. 107–108.

2. Briggs, *Victorian Cities*, p. 86; Tocqueville, *Journeys*, pp. 106–107. For further discussion of reactions to Manchester, see C. F. Chadwick, "The Face of the Industrial City: Two Looks at Manchester," in *The Victorian City: Images and Realities*, ed. H. J. Dyos and Michael Wolff (London, 1973), I, 247–56; and Steven Marcus, *Engels, Manchester, and the Working Class* (New York, 1974), esp. pp. 28–66.

3. N. Scott Momaday, quoted in William Stott, *Documentary Expression and Thirties America* (New York, 1973), p. 47. For an illuminating discussion of the character and aesthetic of social documentary generally, see Stott, pp. 3–63.

4. Charles Dickens, *Hard Times* (London, 1854), pp. 26–27. For Dick-

ens's reactions to his first visit to Manchester in 1838, see Marcus, *Engels*, pp. 30–32.

5. Briggs, *Victorian Cities*, pp. 86–92; Rev. R. Parkinson, quoted in Briggs, p. 111.

6. C. Edwards Lester, *The Glory and the Shame of England*, 1850 ed. (New York, 1850), I, 47. I draw the phrase "emotionally validating image" from Donald Fleming, "Roots of the New Conservation Movement," *Perspectives in American History*, 6 (1972), 52.

7. Benjamin Silliman, *A Journal of Travels in England, Holland and Scotland, and of Two Passages over the Atlantic, in the Years 1805 and 1806 . . .*, 3d. ed. (New Haven, 1820), I, 102, 108.

8. [James Kirke Paulding], *A Sketch of Old England, by a New-England Man* (New York, 1822), I, 144, 149.

9. Lester, *The Glory and the Shame*, I, 207, 139–212, esp. p. 175.

10. Henry Colman, *European Life and Manners; in Familiar Letters to Friends* (Boston, 1850), I, 103.

11. The statistics gathered by Patrick Colquhoun were frequently cited on this point; see his *Treatise on Indigence . . .* (London, 1806), pp. 272–73.

12. Charles L. Sanford, *The Quest for Paradise: Europe and the American Moral Imagination* (Urbana, Ill., 1961), pp. 160–72; Joseph Stancliffe Davis, *Essays in the Earlier History of American Corporations* (Cambridge, Mass., 1917), I, 376; Samuel Rezneck, "The Rise and Early Development of the Industrial Consciousness in the United States, 1760–1830," *Journal of Economic and Business History*, 4 (August 1932), 797.

13. On geographical mobility in nineteenth-century America, see Stephan Thernstrom and Peter R. Knights, "Men in Motion: Some Data and Speculations about Urban Population Mobility in Nineteenth-Century America," in *Anonymous Americans: Explorations in Nineteenth-Century Social History*, ed. Tamara K. Hareven (Englewood Cliffs, N.J., 1971), pp. 17–47.

14. Stanley M. Elkins, *Slavery: A Problem in American Institutional and Intellectual Life* (Chicago, 1959) pp. 27–37; David Donald, "An Excess of Democracy: The American Civil War and the Social Process," in *Lincoln Reconsidered: Essays on the Civil War Era*, 2d ed., Vintage Books (New York, n.d.), pp. 209–35.

15. See for example Michael B. Katz, *The Irony of Early School Reform: Educational Innovation in Mid-Nineteenth Century Massachusetts* (Cambridge, Mass., 1968); Neil Harris, *The Artist in American Society: The Formative Years, 1790–1860* (New York, 1966); Charles I. Foster, *An Errand of Mercy: The Evangelical United Front, 1790–1837*; and David J. Rothman, *The Discovery of the Asylum: Social Order and Disorder in the New Republic* (Boston, 1971).

16. Rothman, *Discovery of the Asylum*, esp. pp. 82–84, 105–108, 133–146, 187–95, 202–205, 207–16; Erving Goffman, *Asylums: Essays on the Social Situation of Mental Patients and Other Inmates*, Anchor Books (Garden City, N.Y., 1961), p. xiii.

17. See Vera Shlakman, *Economic History of a Factory Town: A Study of Chicopee, Massachusetts*, Smith College Studies in History, 20 (October 1934–July 1935), 39–42. On Holyoke, see Constance M. Green,

Holyoke, Massachusetts (New Haven, 1939); on Lawrence, see Donald B. Cole, *Immigrant City: Lawrence, Massachusetts, 1845–1921* (Chapel Hill, 1963).

18. The most useful secondary accounts of nineteenth-century Lowell are: Norman Ware, *The Industrial Worker, 1840–1860: The Reaction of American Industrial Society to the Advance of the Industrial Revolution* (Boston, 1924); Caroline F. Ware, *The Early New England Cotton Manufacture: A Study in Industrial Beginnings* (Boston, 1931); John Coolidge, *Mill and Mansion: A Study of Architecture and Society in Lowell, Massachusetts, 1820–1865* (New York, 1942); and Hannah Josephson, *The Golden Threads: New England's Mill Girls and Magnates* (New York, 1949).

19. In the face of inconclusive evidence, scholarly opinion is divided as to whether F. C. Lowell visited Robert Owen's New Lanark during his critical period of thought and study of manufacturing in 1810–12. Hannah Josephson and Charles Sanford believe that he did; John Coolidge and Ferris Greenslet discount it. However, Lowell certainly was aware of Owen's community since Nathan Appleton visited it during the time he and Lowell were formulating their project in the fall of 1810, though neither man subsequently made public reference to it. See Appleton's Journal for October 16, 1810, Nathan Appleton Papers, Massachusetts Historical Society, Boston. For Robert Owen's critical reaction to Lowell thirty-five years later, see *Voice of Industry*, January 2, 1846, p. 1.

20. Ferris Greenslet, *The Lowells and Their Seven Worlds* (Boston, 1946), p. 156; Nathan Appleton, *Introduction of the Power Loom, and Origin of Lowell* (Lowell, 1858), p. 8; Josephson, *Golden Threads*, pp. 25–26.

21. Appleton, *Introduction*, p. 15; David J. Jeremy, "Innovation in American Textile Technology during the Early 19th Century," *Technology and Culture*, 14 (January 1973), 40–76.

22. Josephson, *Golden Threads*, pp. 29–31; Daniel Webster, *Writings and Speeches*, National ed. (Boston, 1903), XIV, 43, 45. See Clarence Mondale, "Daniel Webster and Technology," *American Quarterly*, 14 (Spring 1962), 37–47; and Leo Marx, *The Machine in the Garden: Technology and the Pastoral Ideal in America* (New York, 1964), pp. 209–20.

23. Appleton, *Introduction*, p. 15.

24. On the transition of women's work from home to factory, see Edith Abbott, *Women in Industry: A Study in American Economic History* (New York, 1924), pp. 48–62; Percy W. Bidwell, "The Agricultural Revolution in New England," *American Historical Review*, 26 (July 1921), 693–96; and Joan W. Scott and Louise A. Tilly, "Women's Work and the Family in Nineteenth-Century Europe," *Comparative Studies in Society and History*, 17 (January 1975), 36–64.

25. John A. Lowell, "Patrick Tracy Jackson," in *Lives of American Merchants*, ed. Freeman Hunt, I (New York, 1856), 565; Mathew Carey, quoted in Abbott, *Women in Industry*, p. 58.

26. H. M. Gitelman, "The Waltham System and the Coming of the Irish," *Labor History*, 8 (Fall 1967), 228–32.

27. Theodore Parker, "The Mercantile Classes," in *Works*, Centenary

ed., X (Boston, 1907), 17; Robert C. Winthrop, *Memoir of the Hon. Nathan Appleton, LL.D.* (Boston, 1861), pp. 60–61; Freeman Hunt, "Amos Lawrence," *Lives of American Merchants,* II (New York, 1858), 287–309; *Extracts from the Diary and Correspondence of the Late Amos Lawrence,* ed. William R. Lawrence (Boston, 1855), p. 202; Edward Weeks, *The Lowells and Their Institute* (Boston, 1966), pp. 26–27. On this general subject see Paul Goodman, "Ethics and Enterprise: The Values of a Boston Elite, 1800–1860," *American Quarterly,* 18 (Fall 1966), 437–51.

28. Victor S. Clark, *History of Manufactures in the United States,* 1929 ed. (New York, 1929), I, 183, 188–89.

29. Appleton, *Introduction,* p. 15.

30. Coolidge, *Mill and Mansion,* pp. 32, 69–70; Bryant Franklin Tolles, Jr., "Textile Mill Architecture in East Central New England: An Analysis of Pre-Civil War Design," *Essex Institute Historical Collections,* 107 (July 1971), 238–39; Rothman, *Discovery of the Asylum,* pp. 107, 153, 181, 191, 226–28.

31. Harriet H. Robinson, *Loom and Spindle, or Life Among the Early Mill Girls* (New York, 1898), pp. 13–15; Coolidge, *Mill and Mansion,* pp. 32–41.

32. George S. White, *Memoir of Samuel Slater ... with Remarks on the Moral Influence of Manufactories in the United States,* 2d ed. (Philadelphia, 1836), pp. 126, 137–38; *Voice of Industry,* March 20, 1846, p. 3. For a defense of work as a means of social control in the late nineteenth century, see [Jonathan Baxter Harrison], "Study of a New England Factory Town," *Atlantic Monthly,* 43 (June 1879), 703. Cf. the stress on work in penitentiaries of the ante-bellum period, discussed in Rothman, *Discovery of the Asylum,* pp. 103–104.

33. Rules of Lawrence Manufacturing Company, Lowell, Mass., quoted in Michel Chevalier, *Society, Manners, and Politics in the United States,* ed. John William Ward, trans. after the T. G. Bradford ed., Cornell Paperback (Ithaca, 1969), p. 140. See also Sidney Pollard, "Factory Discipline in the Industrial Revolution," *Economic History Review,* 2d ser., 16 (December 1963), 254–71; E. P. Thompson, "Time, Work-Discipline, and Industrial Capitalism," *Past and Present,* No. 38 (December 1967), 56–97.

34. Henry A. Miles, *Lowell, As It Was, and As It Is* (Boston, 1845), pp. 65–66, 132–40.

35. Miles, *Lowell,* pp. 144–45. Lowell was not the only manufacturing community which thus relied upon the power of public opinion. See the report of Smith Wilkinson, director of Conger's Mills, Pomfret, Conn., in White's *Memoir of Samuel Slater,* p. 127. As Caroline Ware has pointed out, while English manufactures hired freely from the poorhouses, even the smaller "family" type of mills of southern Massachusetts, Rhode Island, and Connecticut attempted to keep at least a pretense of high moral tone in their communities and advertised that only workers of active and industrious character need apply. C. Ware, *Early New England Cotton Manufacture,* pp. 202–203.

36. C. Ware, *Early New England Cotton Manufacture,* pp. 217, 219–20; Robinson, *Loom and Spindle,* p. 77; "Tales of Factory Life," *Lowell Offering,* new ser., 1 (1841), 65–68; "Prejudice Against Labor," *Lowell Offering,*

new ser., 1 (1841), pp. 136–45; *Lowell Offering*, new ser., 4 (September 1844), 257–58.

37. Coolidge, *Mill and Mansion*, p. 139.

38. Robinson, *Loom and Spindle*, p. 2; Lucy Larcom, "Among Lowell Mill-Girls: A Reminiscence," *Atlantic Monthly*, 48 (November 1881), 600; Lucy Larcom, *A New England Girlhood, Outlined from Memory*, Corinth Books (New York, 1961), pp. 163–64.

39. Larcom, *A New England Girlhood*, pp. 154, 175–76, 183; "Editorial," *Lowell Offering*, new ser., 3 (April 1843), 164. On "removal activities" to pass the time in other total institutions, see Goffman, "On the Characteristics of Total Institutions," in *Asylums*, pp. 68–69.

40. Josephson, *Golden Threads*, p. 86. On the *Lowell Offering* and other factory magazines of the period, see Bertha Monica Stearns, "Early Factory Magazines in New England: The *Lowell Offering* and Its Contemporaries," *Journal of Economic and Business History*, 2 (August 1930), 685–705.

41. "Editorial," *Lowell Offering*, new ser., 5 (January 1845), 22; [Eliza Jane Cate], "Duties and Rights of Mill Girls," *New England Offering*, No. 5 (August 1848), 101.

42. "The Spirit of Discontent," *Lowell Offering*, new ser., 1 (1841), 112; "A Dialogue," *Operatives' Magazine*, No. 3 (June 1841), 37; "Editorial," *Lowell Offering*, new ser., 5 (March 1845), 72.

43. Henry Colman, "Lowell, Massachusetts," *American Railroad Journal* [*Railway Mechanical Engineer*], 5 (April 2, 1836), 197–98.

44. Walton Felch, *The Manufacturer's Pocket-Piece; or the Cotton-Mill Moralized* (Medway, Mass., 1816), p. 14.

45. See Edmund Morgan, ed., *Puritan Political Ideas, 1558–1794* (Indianapolis, 1965), pp. xv–xx.

46. Josephson, *Golden Threads*, pp. 58–61; William Lawrence, *Life of Amos A. Lawrence* (Boston, 1888), p. 18; Charles Cowley, *Illustrated History of Lowell*, revised ed. (Boston, 1868), pp. 79–80, 90.

47. On this attempt, which culminated in the log-cabin campaign of 1840, see John William Ward, *Andrew Jackson, Symbol for an Age* (New York, 1955), chapter 5.

48. David Crockett, *An Account of Col. Crockett's Tour to the North and Down East, in the Year of Our Lord One Thousand Eight Hundred and Thirty-Four* (Philadelphia, 1835), pp. 92–93.

49. Edward Everett, *Orations and Speeches on Various Occasions*, 6th ed. (Boston, 1860), II, 65–66, 52, 63.

50. A valuable study in this connection is Marvin Fisher, *Workshops in the Wilderness: The European Response to American Industrialization, 1830–1860* (New York, 1967).

51. J. S. Buckingham, *The Eastern and Western States of America* (London, 1842), p. 295; Fredrika Bremer, *The Homes of the New World: Impressions of America*, trans. Mary Howitt (New York, 1853), I, 209.

· 52. Alexander Mackay, *The Western World; or, Travels in the United States in 1846–47* (Philadelphia, 1849), p. 262; Charles Dickens, *American Notes for General Circulation* (London, 1842), pp. 152–54; Chevalier, *Society, Manners, and Politics*, pp. 133–34.

53. On institutional displays in total institutions, see Goffman, *Asylums*, pp. 93–112.

54. See William Scoresby, *American Factories and Their Female Operatives*, American ed. (Boston, 1845).

55. Dickens, *American Notes*, pp. 160–61.

56. Harriet Martineau, "The Lowell Offering," *Littell's Living Age*, 2 (September 28, 1844), 502; Harriet Martineau, *Society in America* (London, 1837), II, 354–55. Cf. Patrick Tracy Jackson on the wisdom of a protective tariff for cotton manufactures: "The village steeple is an unfailing companion to the waterwheel, and the liberal professions, and all the arts which minister to the wants and comforts of man, find their best remuneration amidst the population which the enlightened policy of the government has gathered around it." Friends of Domestic Industry, *Report on Manufactures of Cotton*, Convention at the City of New York, October 26, 1831 (Baltimore, 1831–32), p. 111.

57. Chevalier, *Society, Manners, and Politics*, p. 139; Buckingham, *Eastern and Western States*, p. 295.

58. See Fisher, *Workshops in the Wilderness*, p. 41; Marx, *Machine in the Garden*, p. 71.

59. On the chief spokesmen of this movement in the 1830s and their ideology, see Edward Pessen, *Most Uncommon Jacksonians: The Radical Leaders of the Early Labor Movement* (Albany, N.Y., 1967).

60. On Luther's career, see Louis Hartz, "Seth Luther: The Story of a Working-Class Rebel," *New England Quarterly*, 13 (September 1940), 401–18; Carl Gersuny, "Seth Luther—The Road from Chepachet," *Rhode Island History*, 33 (May 1974), 47–55; and Pessen, *Most Uncommon Jacksonians*, pp. 87–90.

61. Seth Luther, *An Address to the Working Men of New England, on the State of Education, and on the Condition of the Producing Classes in Europe and America*, 2d ed. (New York, 1833), pp. 35, 12–13, 14n.

62. Luther, *Address to Working Men*, p. 17; quotation from John Quincy Adams.

63. That American factory operatives worked longer hours than their English counterparts was confirmed in George Wallis, "Report on Manufactures," *General Report of the British Commissioners Appointed to Attend the Exhibition of Industry in the City of New York* (London, 1854), pp. 4, 9.

64. Luther, *Address to Working Men*, pp. 7–8, 29.

65. Luther, *Address to Working Men*, p. 27.

66. Seth Luther, *An Address on the Origin and Progress of Avarice, and its Deleterious Effects on Human Happiness, with a Proposed Remedy for the Countless Evils Resulting from an Inordinate Desire for Wealth* (Boston, 1834), p. 16.

67. Luther, *Address to Working Men*, p. 32.

68. Luther, *Address on Avarice*, p. 40; Hartz, "Seth Luther," p. 410; Gersuny, "Seth Luther," p. 47.

69. On Douglas's career, see Pessen, *Most Uncommon Jacksonians*, pp. 90–91.

70. John R. Commons *et al.*, eds., *A Documentary History of American Industrial Society* (Cleveland, 1910), VI, 217–18.

71. Orestes A. Brownson, "The Laboring Classes," *Boston Quarterly Review*, 3 (July, October 1840), 472–73, 364.

72. Brownson, "Laboring Classes, p. 508.

73. For the response to "The Laboring Classes," see Arthur M. Schlesinger, Jr., *A Pilgrim's Progress: Orestes A. Brownson*, revised ed. (Boston, 1966), pp. 100–11.

74. Elisha Bartlett, *A Vindication of the Character and Condition of the Females Employed in the Lowell Mills, Against the Charges Contained in the Boston Times, and the Boston Quarterly Review* (Lowell, 1841).

75. *Corporations and Operatives: Being an Exposition of the Condition of Factory Operatives, and a Review of the "Vindication," by Elisha Bartlett, M.D.* (Lowell, 1842), pp. 47–71.

76. John B. Andrews, "History of Women in Trade Unions, 1825 through the Knights of Labor," in John B. Andrews and W. D. P. Bliss, *History of Women in Trade Unions* (Washington, D.C., 1911), p. 27–29; quotation from *The Man*, March 20, 1834.

77. Andrews, "History of Women," p. 30.

78. For other examples of workers' appeals to the Declaration of Independence and the American Revolution in this period, see Herbert G. Gutman, "Work, Culture, and Society in Industrializing America, 1815–1919," *American Historical Review*, 78 (June 1973), 568. On the general subject of the Declaration of Independence and American radicalism in the eighteenth and nineteenth centuries, see Staughton Lynd, *Intellectual Origins of American Radicalism* (New York, 1968).

79. *Diary and Correspondence of Amos Lawrence*, pp. 301, 166; Hunt, *Lives of American Merchants*, II, 244.

80. C. Ware, *Early New England Cotton Manufacture*, p. 276.

81. George M. Fredrickson and Christopher Lasch, "Resistance to Slavery," *Civil War History*, 13 (December 1967), 317*n*.

82. Josephson, *Golden Threads*, pp. 194–98, 202–203. On house organs in total institutions, see Goffman, *Asylums*, pp. 95–96.

83. "Editor's Table," *New England Offering*, No. 4 (July 1848), 95; Lucy Larcom, *An Idyl of Work* (Boston, 1875), pp. 188–89; "Conclusion of the Volume," *Lowell Offering*, new ser., 1 (1841), 377.

84. Stephan Thernstrom, "Working-Class Social Mobility in Industrial America," in *Essays in Theory and History: An Approach to the Social Sciences*, ed. Melvin Richter (Cambridge, Mass., 1970), 225.

85. Pessen, *Most Uncommon Jacksonians*, p. 49; National Trades' Union, 1839 Convention, Report of the Committee on Female Labor, in Commons *et al.*, eds., *Documentary History*, VI, 281–83.

86. Josephson, *Golden Threads*, pp. 199–202; *Voice of Industry*, May 29, 1845, p. 3; *Voice of Industry*, September 4, 1845, p. 2; "The Evils of Factory Life," *Factory Tracts*, No. 1 (1845), pp. 3–4.

87. Thompson, "Time, Work-Discipline, and Industrial Capitalism," p. 86.

88. For a somewhat different view of this subject, see Pessen, *Most Uncommon Jacksonians*, pp. 191–95.

89. On politics as symbolic action, see Joseph R. Gusfield, *Symbolic Crusade: Status Politics and the American Temperance Movement* (Urbana, Ill., 1963), esp. pp. 167–72.

90. Commons *et al.*, eds., *Documentary History*, VIII, 140–41.

91. Commons *et al.*, eds., *Documentary History*, VIII, 150–51; Cowley, *Illustrated History of Lowell*, p. 149.

92. Miles, *Lowell*, pp. 116–27, 215–16; Lawrence to Henry Whiting, November 1849, *Diary and Correspondence of Amos Lawrence*, p. 274.

93. Thomas G. Cary, *Profits on Manufactures at Lowell* (Boston, 1845); John Aiken, *Labor and Wages, At Home and Abroad* (Lowell, 1849), pp. 15, 16. See also Irvin G. Wyllie, *The Self-Made Man in America: The Myth of Rags to Riches* (New Brunswick, N.J., 1954).

94. Nathan Appleton, "Labor—Its Relations in the United States and Europe, Compared," [Hunt's] *Merchants' Magazine and Commerical Review*, 11 (September 1844), 217–23. See also *Correspondence between Nathan Appleton and John A. Lowell in Relation to the Early History of the City of Lowell* (Boston, 1848); and Appleton, *Introduction*.

95. Nathan Appleton, "Sketches of Autobiography," quoted in Winthrop, *Memoir of Hon. Nathan Appleton*, pp. 12–13. In fact, Appleton's father was financially well off by the standards of a farming community in the late eighteenth century, as well as socially and politically prominent. See Francis W. Gregory, "Nathan Appleton, Yankee Merchant (1779–1861)" (Ph.D. diss., Radcliffe College, 1949), pp. 13–14.

96. On this point see Stephan Thernstrom, *Poverty and Progress: Social Mobility in a Nineteenth Century City* (Cambridge, Mass., 1964), pp. 57–79. Industrialists and their spokesmen repeated this message throughout the nineteenth century until in 1889 Richard T. Ely finally protested: "If you tell a single concrete workman on the Baltimore and Ohio Railroad that he may yet be president of the company, it is not demonstrable that you have told him what is not true, although it is within bounds to say that he is far more likely to be killed by a stroke of lightning." Quoted in R. Richard Wohl, "The 'Rags to Riches Story': An Episode of Secular Idealism," in *Class, Status and Power: Social Stratification in Comparative Perspective*, ed. Reinhard Bendix and Seymour Martin Lipset, 2d ed. (New York, 1966), p. 504.

97. Paul F. McGouldrick, *New England Textiles in the Nineteenth Century: Profits and Investment* (Cambridge, Mass., 1968). See also Lance Edwin Davis, "Stock Ownership in the Early New England Textile Industry," *Business History*, 32 (Summer 1958), 204–22; and Robert G. Layer, *Earnings of Cotton Mill Operatives, 1825–1914* (Cambridge, Mass., 1955).

98. McGouldrick, *New England Textiles*, p. 20; C. Ware, *Early New England Cotton Manufacture*, pp. 111–12; Percy Wells Bidwell, "Population Growth in Southern New England, 1810–60," *Publications of the American Statistical Association*, new ser., 15 (December 1917), 835; Josephson, *Golden Threads*, pp. 207, 220, 294.

99. Oscar Handlin, *Boston's Immigrants: A Study in Acculturation*, revised ed. (Cambridge, Mass., 1959), pp. 35–87.

100. C. Ware, *Early New England Cotton Manufacture*, pp. 228–31, 259–60; Abbott, *Women in Industry*, p. 139; Josephson, *Golden Threads*,

p. 296; Gitelman, "The Waltham System and the Coming of the Irish," pp. 240–53.

    101. See for example, Miles, *Lowell*, pp. 75–76, 130.

    102. Charles Cowley, *A Hand Book of Business in Lowell, with a History of the City* (Lowell, 1856), pp. 162–64. Cf. Aiken, *Labor and Wages*, p. 10.

    103. Rothman, *Discovery of the Asylum*, pp. 237–64; Katz, *Irony of Early School Reform*, pp. 180–81; Gerald N. Grob, "Mental Illness: Indigency and Welfare: The Mental Hospital in Nineteenth-Century America," in *Anonymous Americans*, ed. Hareven, pp. 260–71.

3. *Technology and Imaginative Freedom: R. W. Emerson*

    1. On this general subject see Howard Mumford Jones, *O Strange New World: American Culture: The Formative Years* (New York, 1964), esp. pp. 312–50; and Russel Blaine Nye, *The Cultural Life of the New Nation, 1776–1830* (New York, 1960).

    2. John Jersey Mawyer, *An Oration Delivered to French Calvinists, The Fifth of July, 1819* (Charleston, 1819), p. 17, quoted in Merle Curti, *The Roots of American Loyalty* (New York, 1946), p. 61; *Beloit College Monthly*, 4 (October 1856), p. 28, quoted in Curti, *Roots*, p. 62; *The Works of Walt Whitman*, "Deathbed" ed. (1891–92, rpt. New York, 1968), II, 209. See Benjamin T. Spencer, *The Quest for Nationality: An American Literary Campaign* (Syracuse, 1957).

    3. A phrase Mark Schorer has applied to William Blake; quoted in M. H. Abrams, "English Romanticism: The Spirit of the Age," in *Romanticism Reconsidered: Selected Papers from the English Institute*, ed. Northrop Frye (New York, 1963), p. 44.

    4. Abrams, "English Romanticism," p. 31; Abrams also notes the impact of the American Revolution on English Romantics.

    5. *The Complete Works of Ralph Waldo Emerson*, ed. Edward Waldo Emerson (Centenary ed., Boston, 1903–1904), I, 110, 144; Emerson to Edward Bliss Emerson, May 31, 1834, *The Letters of Ralph Waldo Emerson*, ed. Ralph L. Rusk (New York, 1939), I, 413. See René Wellek, "Emerson and German Philosophy," *New England Quarterly*, 16 (March 1943), 41–62.

    6. John Dewey, *Characters and Events: Popular Essays in Social and Political Philosophy*, ed. Joseph Ratner (New York, 1929), I, 75–76.

    7. Emerson to John Boynton Hill, July 3, 1822, *Letters*, I, 121; Perry Miller, "Emersonian Genius and the American Democracy," in *Nature's Nation* (Cambridge, Mass., 1967), p. 164; *The Journals and Miscellaneous Notebooks of Ralph Waldo Emerson*, ed. William H. Gilman *et al.*, 10 vols. to date (Cambridge, Mass., 1960–73), VI, 224; IX, 269; IV, 357. I have followed this edition, omitting Emerson's own cancellations and following his corrections, for all references to Emerson's journals to 1848; thereafter, I

have had to rely on the older edition, *The Journals of Ralph Waldo Emerson*, ed. Edward Waldo Emerson and Waldo Emerson Forbes, Centenary ed. (Boston, 1909-14).

8. William Charvat, *Emerson's American Lecture Engagements: A Chronological List* (New York, 1961).

9. *Works*, VI, 49. On this point, see Harold Kaplan, *Democratic Humanism and American Literature* (Chicago, 1972), p. 25; Stephen E. Whicher, *Freedom and Fate: An Inner Life of Ralph Waldo Emerson* (Philadelphia, 1953); and Harold Bloom, "Emerson: The Glory and the Sorrows of American Romanticism," *Virginia Quarterly Review*, 47 (Autumn 1971), 546-63.

10. *Works*, XII, 11; IV, 62.

11. See for example George Santayana, "The Genteel Tradition in American Philosophy," in *Winds of Doctrine: Studies in Contemporary Opinion* (New York, 1913), pp. 186-215; Lionel Trilling, "Reality in America," in *The Liberal Tradition: Essays on Literature and Society*, Anchor Books (Garden City, N.Y., 1953), pp. 15-21; R. W. B. Lewis, *The American Adam: Innocence, Tragedy and Tradition in the Nineteenth Century* (Chicago, 1955), pp. 1-10; Kaplan, *Democratic Humanism and American Literature*, pp. 27-28, 37-41.

12. *Journals and Miscellaneous Notebooks*, IV, 226.

13. On this point see Sherman Paul, *Emerson's Angle of Vision: Man and Nature in American Experience* (Cambridge, Mass., 1952), esp. pp. 71-102.

14. *Journals and Miscellaneous Notebooks*, IV, 82.

15. Mark Twain, "Old Times on the Mississippi," *Atlantic Monthly*, 35 (March 1875), 288-89; this writing was later incorporated into *Life on the Mississippi* (Boston, 1883). Charles A. Lindbergh, *The Spirit of St. Louis* (New York, 1953), p. 249.

16. *Journals and Miscellaneous Notebooks*, IV, 296; Paul, *Emerson's Angle of Vision*, pp. 73-84.

17. *Works*, I, 13-14, 295-96; *Journals and Miscellaneous Notebooks*, VIII, 397; Perry Miller, ed., *The Transcendentalists* (Cambridge, Mass., 1950), p. 248.

18. *Works*, I, 41.

19. On this shift in Emerson's thought, see especially Whicher, *Freedom and Fate*, and Jonathan Bishop, *Emerson on the Soul* (Cambridge, Mass., 1964), pp. 165-227. Bishop particularly connects the change in Emerson's thought with the death of his first-born son, Waldo, in 1842.

20. *Journals and Miscellaneous Notebooks*, VII, 342, 268; Ralph L. Rusk, *The Life of Ralph Waldo Emerson* (New York, 1949), p. 285.

21. *Works*, III, 6, 18-19; Melville's comment is quoted in F. O. Matthiessen, *American Renaissance: Art and Expression in the Age of Emerson and Whitman* (New York, 1941), pp. 54-55.

22. *Works*, III, 37.

23. *Works*, I, 451-55.

24. *Works*, I, 365, 370.

25. *Works*, I, 373-75; Whicher, *Freedom and Fate*, p. 140.

26. *Journals and Miscellaneous Notebooks*, X, 102-103.

27. *Journals and Miscellaneous Notebooks*, IV, 81; X, 239; Emerson to Lidian Emerson, Dec. 1, 1847 and Mar. 10, 1848, *Letters*, III, 442; IV, 34–35.

28. *Works*, V, 220, 255, 35, 204, 251, 98, 95.

29. *Works*, V, 159.

30. *Works*, V, 166–67. See Lewis Mumford, *Technics and Civilization* (New York, 1934), pp. 250–55. On the origins of the reaper in the imitation of the hand and its development, see Siegfried Giedion, *Mechanization Takes Command: A Contribution to Anonymous History* (New York, 1948), pp. 150–53.

31. *Works*, I, 240–41; V, 103; F. B. Sanborn, ed., "The Emerson-Thoreau Correspondence," *Atlantic Monthly*, 69 (June 1892), 742.

32. *Works*, V, 215. On the theory of history in *English Traits*, see Philip L. Nicoloff, *Emerson on Race and History: An Examination of English Traits* (New York, 1961).

33. *Works*, V, 275–76, 35.

34. "The Anglo-American," quoted in Michael H. Cowan, *City of the West: Emerson, America, and Urban Metaphor* (New Haven, 1967), p. 49; *Journals and Miscellaneous Notebooks*, X, 358.

35. Holman Hamilton, *Prologue to Conflict: The Crisis and Compromise of 1850* (Lexington, 1964).

36. Emerson's great-grandfather, the Reverend Joseph Emerson, reportedly "prayed every night that none of his descendants might ever be rich"; James Eliot Cabot, *A Memoir of Ralph Waldo Emerson* (Boston, 1887), I, 9. See Perry Miller, "Sinners in the Hands of a Benevolent God," in *Nature's Nation*, pp. 279–89.

37. *Works*, IX, 78; XI, 183–86.

38. *Journals*, VIII, 445; IX, 498.

39. *Journals*, VII, 518.

40. *Works*, VI, 94.

41. *Works*, I, 192; VI, 17, 32; VIII, 221; VI, 240.

42. Whicher, *Freedom and Fate*, pp. 163–64; *Journals*, VIII, 434, 481. See also *Journals and Miscellaneous Notebooks*, X, 300; *Works*, VIII, 227; VI, 68.

4. The Aesthetics of Machinery

1. Walt Whitman, "Song of the Exposition," *Leaves of Grass*, "Deathbed" ed. (Philadelphia, 1891–92), p. 159; *The Complete Works of Ralph Waldo Emerson*, ed. Edward Waldo Emerson, Centenary ed. (Boston, 1903–1904), II, 368.

2. "Miscellaneous," *Scientific American*, 6 (May 31, 1851), 290; "Letters from Susan," *Lowell Offering*, new ser., 4 (August 1844), 238; "Our Manufactures," *Scientific American*, 6 (December 7, 1850), 93.

3. Walt Whitman, *The Gathering of the Forces*, ed. Cleveland Rogers and John Black (New York, 1920), II, 210; James Richardson, "Traveling

by Telegraph: Northward to Niagara," *Scribner's Monthly*, 4 (May–June 1872), 22; Nathaniel Hawthorne, *English Notebooks*, ed. Randall Stewart (New York, 1941), p. 279.

4. Alexis de Tocqueville, *Democracy in America*, ed. J. P. Mayer, trans. George Lawrence, Anchor Books (Garden City, N.Y., 1969), p. 465.

5. Quoted in Constance Rourke, *The Roots of American Culture and Other Essays*, ed. Van Wyck Brooks (New York, 1942), p. 3. See also Neil Harris, *The Artist in American Society: The Formative Years 1790–1860* (New York, 1966), pp. 33–41.

6. Wendell D. Garrett, "John Adams and the Limited Role of the Fine Arts," *Winterthur Portfolio*, 1 (1964), 243–55.

7. See particularly: Harris, *Artist in American Society*; Lillian B. Miller, *Patrons and Patriotism: The Encouragement of the Fine Arts in the United States, 1790–1860* (Chicago, 1966); and Alan Gowans, "Taste and Ideology: Principles for a New American Art History," in *The Shaping of Art and Architecture in Nineteenth-Century America* (New York, 1972), pp. 156–87.

8. On Greenough and Emerson as prophets of functionalism, see Robert B. Shaffer, "Emerson and His Circle: Advocates of Functionalism," *Journal of the Society of Architectural Historians*, 7 (July–December 1948), 17–20; Erle Loran, "Introduction" to Horatio Greenough, *Form and Function: Remarks on Art, Design, and Architecture*, ed. Harold A. Small (Berkeley, 1947), pp. xiii–xxi; Charles R. Metzger, *Emerson and Greenough: Transcendental Pioneers of an American Esthetic* (Berkeley, 1954); Nathalia Wright, "Ralph Waldo Emerson and Horatio Greenough," *Harvard Library Bulletin*, 12 (Winter 1958), 91–116; and James Marston Fitch, *Architecture and the Esthetics of Plenty* (New York, 1961), chapter 4, "Horatio Greenough, Yankee Functionalist," pp. 46–64. For the sources of functionalist aesthetics generally, see Edward Robert De Zurko, *Origins of Functionalist Theory* (New York, 1957); for Greenough's debt to neoclassical aesthetics, see Sylvia E. Crane, "The Aesthetics of Horatio Greenough in Perspective," *Journal of Aesthetics and Art Criticism*, 24 (Spring 1966), 415–27; for the way in which various ante-bellum styles were defended as organic and functional architecture, see James Early, *Romanticism and American Architecture* (New York, 1965), pp. 70–71, 84–111, 155–57. As Janet Malcolm has observed, the statements of Greenough, Andrew Jackson Downing, and other nineteenth-century aestheticians must be judged in the light of their actual productions rather than twentieth-century preconceptions: "Functionalism is often in the mouth of the designer rather than the eye of the beholder." "On and Off the Avenue," *New Yorker*, 49 (June 9, 1973), 90.

9. "American Architecture," in Henry T. Tuckerman, *A Memorial of Horatio Greenough* (1853; rpt. New York, 1968), p. 123.

10. Harris, *Artist in American Society*, pp. 56–64. See also Charles Coleman Sellers, *Charles Willson Peale* (New York, 1969); Sylvia E. Crane, *White Silence: Greenough, Powers, and Crawford, American Sculptors in Nineteenth-Century Italy* (Coral Gables, 1972), pp. 257–62; Henry W. Dickinson, *Robert Fulton, Engineer and Artist* (1913; rpt. Freeport, N.Y., 1971); Oliver W. Larkin, *Samuel F. B. Morse and American Democratic*

Art (Boston, 1954); and Jean Lipman, *Rufus Porter: Yankee Pioneer* (New York, 1968).

11. Edward Everett, *Orations and Speeches on Various Occasions*, 6th ed. (Boston, 1860), II, 247; "The Poetry of Discovery," *Scientific American*, 5 (November 24, 1849), 77; "The Steam Engines," *Scientific American*, 9 (September 24, 1853), 14; New York *Illustrated News*, quoted in Charles Hirschfeld, "America on Exhibition: The New York Crystal Palace," *American Quarterly*, 9 (Summer 1957), 114–15.

12. See Monte A. Calvert, *The Mechanical Engineer in America, 1830–1910: Professional Cultures in Conflict* (Baltimore, 1967).

13. On Ewbank's life see Carl W. Mitman's sketch in the *Dictionary of American Biography*.

14. Thomas Ewbank, *The World a Workshop: or, The Physical Relationship of Man to the Earth* (New York, 1855), pp. 27, 141; *Report of the Commissioner of Patents for the Year 1849, Part I, Arts and Manufactures*, 31st Congress, 1st Session, House of Representatives, Executive Document No. 20, p. 488.

15. *Report of Commissioner of Patents*, pp. 492, 488; Thomas Ewbank, "Artists of the Ideal and the Real; or, Poets and Inventors—Revival of an Old Mode of Carving," *Scientific American*, 4 (December 23, 1848), 107.

16. Harris, *Artist in American Society*, pp. 300–16.

17. *Report on Commissioner of Patents*, pp. 492, 488; *DAB*, s.v. Ewbank.

18. "Utility of American Inventions," *Scientific American*, 7 (September 27, 1851), 13; "Modern and Ancient Works," *Scientific American*, 4 (February 10, 1849), 165.

19. *Scientific American* ominously noted a report that novel-reading might lead to insanity. "Novels and Insanity," *Scientific American*, 4 (April 7, 1849), 230.

20. "Physical Science and the Useful Arts in Their Relation to Christian Civilization," *New Englander*, 9 (November 1851), 489–90.

21. John C. Kimball, "Machinery as a Gospel Worker," *Christian Examiner*, 87 (November 1869), 323–24, 327.

22. John A. Kouwenhoven, *Made in America: The Arts in Modern Civilization* (Garden City, N.Y., 1948), pp. 15–53, esp. pp. 30–31; reissued by W. W. Norton under title, *The Arts in Modern American Civilization* (New York, 1967).

23. On this subject see the excellent essay by Monte A. Calvert, "American Technology at World Fairs, 1851–1876" (M.A. thesis, University of Delaware, 1962).

24. Horace Greeley, ed., *Art and Industry as Represented in the Exhibition at the Crystal Palace, New York, 1853–54* (New York, 1853), p. 305.

25. John H. White, Jr., "Grant's Silver Locomotive," Railway and Locomotive Historical Society *Bulletin*, 104 (April 1961), 54–59; "The Paris Exposition," *Scientific American*, 2d ser., 16 (May 18, 1867), 311; "The Paris Exposition–As It Opened," *Scientific American*, 2d ser., 16 (May 11, 1867), 294.

26. [John Anderson], "Extracts from British Reports Referring to the

Exhibits of the United States at the International Exhibition, Vienna, 1873,"
in *Reports of the Commissioners of the United States to the International
Exhibition Held in Vienna, 1873*, ed. Robert H. Thurston (Washington,
D.C., 1876), I, 233; see also III, 247; and "Machine Tools at the Vienna Exhibition," *Engineering*, 16 (August 22, 1873), 148.

27. "The International Exhibition of 1876," *Scientific American Supplement*, 1 (June 17, 1876), 386; "Machine Tools at the Philadelphia Exhibition," *Engineering*, 21 (May 26, 1876), 427–28.

28. Joseph Wickham Roe, *English and American Tool Builders* (New
Haven, 1916), p. 63; Kouwenhoven, *Made in America*, p. 26.

29. W. H. Mayall, *Machines and Perception in Industrial Design* (New
York, 1968), p. 15.

30. Samuel Clegg, *Architecture of Machinery: An Essay on Propriety
of Form and Proportion, With a View to Assist and Improve Design* (London, 1842), p. 2.

31. Roe, *English and American Tool Builders*, p. 248; Kouwenhoven,
Made in America, p. 32; Mayall, *Machines and Perception*, p. 15.

32. Calvert, "American Technology at World Fairs," p. 189.

33. Lewis Mumford, *Technics and Civilization* (New York, 1934), p.
345; Marvin Fisher, *Workshops in the Wilderness: The European Response
to American Industrialization, 1830–1860* (New York, 1967), pp. 149–50,
170.

34. Franz Reuleaux, the German Commissioner General to the Philadelphia Centennial Exposition of 1876, saw a steam engine with the beautifully lettered inscription, "Our Pet." John Maass, *The Glorious Enterprise: The Centennial Exhibition of 1876 and H. J. Schwarzmann, Archi-
tect-in-Chief* (Watkins Glen, N.Y., 1973), p. 110*n*.

35. The iconographic significance of the Columbian press was not lost
on Europeans. T. C. Hansard noted in his *Typographia* (London, 1825): "If
the merits of a machine were to be appreciated wholly by its ornamental
appearance, certainly no other press could enter into competition with 'The
Columbian.' No British-made machinery was ever so lavishly embellished.
We have a somewhat highly-sounding title to begin with; and then, which
way soever our eyes are turned, from head to foot, or foot to head, some
extraordinary features present themselves—on each pillar of the staple a
caduceus of the universal messenger, Hermes—alligators, and other draconic
serpents, emblematize, on the levers, the power of wisdom—then, for *the
balance of power* (we rude barbarians of the old word make mere cast-iron
lumps serve to inforce our notions of the *balance of power*) we see, sur-
mounting the Columbian press, the American eagle with extended wings,
and grasping in his talons Jove's thunderbolts, combined with the olive-
branch of Peace and cornucopia of Plenty, all handsomely bronzed and gilt
resisting and bearing down ALL OTHER POWER!" Quoted in James Moran,
"The Columbian Press," *Journal of the Printing Historical Society*, No. 5
(1969), 1.

36. Lewis Mumford, *Art and Technics* (New York, 1952), p. 80.

37. On "implicit movement" and an expressive aesthetic see David Ca-
rew Huntington, *Art and the Excited Spirit* (Ann Arbor, 1972).

38. See for example, "Characteristics of the International Fair," *Atlantic Monthly*, 38 (October 1876), 358.

39. Philip T. Sandhurst, *et al.*, *The Great Centennial Exhibition Critically Described and Illustrated* (Philadelphia, 1876), pp. 365–68.

40. "Characteristics of the International Fair," p. 358.

41. "Motive Power of the International Exhibition," *Journal of the Franklin Institute*, 103 (January 1877), 4; Charles T. Porter, *Engineering Reminiscences* (New York, 1908), p. 248, cited in Calvert, "American Technology at World Fairs," p. 173.

42. "Opening of the International Exhibition," *Scientific American Supplement*, 1 (May 27, 1876), 338–39; New York *Herald*, May 11, 1876, quoted in Christine Hunter Donaldson, "The Centennial of 1876: The Exposition and Culture for America" (Ph.D. diss., Yale University, 1948), p. 8.

43. Richard Henry Stoddard, ed., *A Century After: Picturesque Glimpses of Philadelphia and Pennsylvania* (Philadelphia, 1876), p. 348; "The Exhibition, *Nation*, 22 (May 18, 1876), 319; Joaquin Miller, "The Great Centennial Fair and Its Future," *The Independent*, 28 (July 13, 1876), 2; [Marietta Holley] *Josiah Allen's Wife as a P.A. and P.I.: Samantha at the Centennial* (Hartford, 1878), p. 508; Sandhurst, *The Great Centennial Exhibition*, p. 361.

44. William Dean Howells, "A Sennight of the Centennial," *Atlantic Monthly*, 38 (July 1876), 96. Like Howells and other visitors, the Philadelphia architect Frank Furness was deeply impressed by the apparent fusion of animal and mechanical attributes in the Corliss engine—a combination which inspired him in his design of the Provident Trust Company building of 1879. Thus American machinery in part repaid its aesthetic debt to architecture. See Neil A. Levine, "The Idea of Frank Furness' Buildings" (M.A. thesis, Yale University, 1967), pp. 76–89.

45. See particularly Bernard Bowron, Leo Marx, and Arnold Rose, "Literature and Covert Culture," *American Quarterly*, 9 (Winter 1957), 377–86; Leo Marx, *The Machine in the Garden: Technology and the Pastoral Ideal in America* (New York, 1964), pp. 207–208; and Fisher, *Workshops in the Wilderness*, pp. 148–71.

46. See, for example, Charles A. Fenton, " 'The Bell-Tower': Melville and Technology," *American Literature*, 28 (May 1951), 219–32; and Marvin Fisher, "Melville's 'Bell-Tower': A Double Thrust," *American Quarterly*, 18 (Summer 1966), 200–207.

47. Edmund Burke, *A Philosophical Enquiry into the Origin of our Ideas of the Sublime and Beautiful*, ed. James T. Boulton, University of Notre Dame Paperback. (Notre Dame, Ind., 1968), pp. 39, 51, 65, 82.

48. See Samuel H. Monk, *The Sublime: A Study of Critical Theories in XVIII-Century England* (New York, 1935). For the development of the technological sublime in England, see Francis D. Klingender, *Art and the Industrial Revolution*, ed. and revised by Arthur Elton (New York, 1968), esp. pp. 83–103.

49. *The Recollections of John Ferguson Weir*, ed. Theodore Sizer (New York, 1957), p. 39.

50. Among the artists who visited Weir's room at the famous Tenth

Street Studios to view his work were Frederick Church, Asher B. Durand, Eastman Johnson, John Kensett, Sanford Gifford, and Worthington Whittredge. When the painting was displayed at the National Academy of Design in 1866, it caused Weir's unanimous election as an Academician. At the Paris Exposition of 1867, it continued to receive critical acclaim. *Recollections of John Ferguson Weir*, pp. 52–58.

51. Kenneth Clark, *The Nude: A Study in Ideal Form* (New York, 1956), p. 187.

52. Klingender, *Art and the Industrial Revolution*, pp. 57–64. A. Pigler lists well over a hundred works from the fifteenth through the nineteenth centuries depicting Venus and Vulcan at the forge and related subjects in *Barockthemen* (Berlin, 1956), II, 20–21, 38, 159–60, 256, 275–78. A more immediate predecessor to Weir's work was Joseph Wright of Derby, who pioneered dramatic, artificially lit, and sublimely charged forge paintings and who influenced Robert Weir's painting, *The Microscope* (1849; Yale University Art Gallery, New Haven). See Benedict Nicolson, "Joseph Wright's Early Subject Pictures," *Burlington Magazine*, 96 (March 1954), 76. J. F. Weir may have also been influenced by the Swedish painter Pehr Hilleström's *In the Foundry: Smiths at Work* (1782; Nationalmuseum, Stockholm), which resembles *The Gun Foundry* in a number of respects: setting, general composition, lighting, workers' attitudes, even the presence of the forge owner and his party of visitors to the right. See Sixten Ronnow, *Pehr Hilleström och hans Bruk-och Bergverksmalningar* (Stockholm, 1920), pp. 159–61. Weir had the opportunity to see engravings of many of these works in the outstanding collection of books and prints owned by his father, Robert Weir. Dorothy Weir Young, *The Life and Letters of J. Alden Weir*, ed. Lawrence W. Chisolm (New Haven, 1960), p. 5.

53. *The Aeneid of Virgil*, trans. John D. Long (Boston, 1879), Bk. VIII, lines 550–60.

54. [George William Curtis], "Editor's Easy Chair," *Harper's Monthly Magazine*, 33 (June 1866), 117.

55. "Darkness," wrote Burke, "is more productive of sublime ideas than light." *Enquiry into the Sublime and Beautiful*, p. 80.

56. J. O. Davidson, "Interior of a Southern Cotton Press by Night," *Harper's Weekly*, 27 (March 24, 1883), 182; S. G. W. Benjamin, "The Works of the Bethlehem Iron Company," *Harper's Weekly*, 35 (March 14, 1891), 194.

57. *The Diary of George Templeton Strong*, ed. Allan Nevins and Milton Halsey Thomas (New York, 1952), I, 108; "The American Woodburning Locomotive," *Scientific American*, 6 (April 3, 1851), 227. Interestingly, both Strong and *Scientific American*'s reporter alluded to an anecdote of a man encountering a train for the first time as he traveled along a road at night. Emerson repeated the tale in his journal in 1835: "Good story Squire Adams told here of countryman travelling between day and sunrise, & seeing the locomotive & its train of cars on the rail road. He saw the smoke & the wheels. His horse was frightened, ran, turned over the wagon & broke it. He crawled to a house for help: They asked him what had happened: He could not well tell what, but that it looked like hell in harness."

The Journals and Miscellaneous Notebooks of Ralph Waldo Emerson, ed. W. H. Gilman *et al.* (Cambridge, Mass., 1960–   ), V, 41. Davy Crockett gave a Southern version of the same story in *An Account of Col. Crockett's Tour to the North and Down East in the Year of Our Lord One Thousand Eight Hundred and Thirty-Four* (Philadelphia, 1835).

58. Walt Whitman, "To a Locomotive in Winter," *Leaves of Grass*, p. 359; Lyonel Feininger to Alfred H. Barr, Jr., quoted in Alfred H. Barr, Jr., "Lyonel Feininger—American Artist," in *Lyonel Feininger—Marsden Hartley*, ed. Dorothy C. Miller and Hudson D. Walker (1944, rpt. New York, 1966), pp. 7–8.

59. See Marx, *Machine in the Garden*, esp. pp. 11–33.

60. Marx, *Machine in the Garden*, pp. 209–14.

61. Everett, *Orations and Speeches*, III, 85–86.

62. "An Excursion on the Baltimore and Ohio Railroad," *The Crayon*, 5 (July 1858), 208–10; "Artists' Excursion over the Baltimore and Ohio Railroad," *Harper's New Monthly Magazine*, 19 (June 1859), 1–19.

63. Richardson, "Traveling by Telegraph," p. 17.

64. LeRoy Ireland, comp., *The Works of George Inness: An Illustrated Catalogue Raisonné* (Austin, 1965), p. 28; George Inness, Jr., *Life, Art, and Letters of George Inness* (New York, 1917), pp. 108–11. The most interesting analyses of this painting are contained in Wolfgang Börn, *American Landscape Painting: An Interpretation* (New Haven, 1948), pp. 156–59; and Marx, *Machine in the Garden*, pp. 220–21.

65. See, for example, *After the Wedding in Warren, Pennsylvania* (c. 1862); Dana Smith, *New Hampshire Panorama* (c. 1860); and Leila T. Bauman, *Geese in Flight* (c. 1870), reproduced in *American Naïve Painting of the Eighteenth and Nineteenth Centuries: 111 Masterpieces from the Collection of Edgar William and Bernice Chrysler Garbisch* (New York, 1969), pls. 101, 95, 99.

66. Walt Whitman, "Passage to India," *Leaves of Grass*, p. 317.

67. This invocation of the famous quotation from Bishop Berkeley's "Verses on the Prospect of Planting Arts and Learning in America" may well derive from Emanuel Leutze's *Westward the Course of Empire Takes Its Way*. Leutze made a sketch for the large fresco executed in the staircase of the Capitol in 1868 (National Collection of Fine Arts, Washington, D.C.). Technology is absent from Leutze's picture, however, while in Currier & Ives's it lies at the very center.

68. Roy King and Burke Davis, *The World of Currier & Ives* (New York [1968]), p. 64. On Mrs. Palmer, see Mary Bartlett Cowdrey, "Fanny Palmer, an American Lithographer," *Prints, Thirteen Illustrated Essays on the Art of the Print*, ed. Carl Zigrosser (New York, 1962), pp. 217–34; and Harry T. Peters, *Currier & Ives: Printmakers to the American People*, I (Garden City, N.Y., 1929), 110–16. Mrs. Palmer's sketches for *Across the Continent* are reproduced in Peters, I, pls. 18–19.

69. On the American West and the myth of the garden, see Henry Nash Smith, *Virgin Land: The American West as Symbol and Myth* (Cambridge, Mass., 1950), book 3.

70. *Journals and Miscellaneous Notebooks*, X, 353.

5. Technology and Utopia

1. Robert H. Wiebe, *The Search for Order, 1877–1920* (New York, 1967), p. xiii.

2. Herbert G. Gutman, "Work, Culture, and Society in Industrializing America, 1815–1919," *American Historical Review*, 78 (June 1973), 555.

3. Andrew Carnegie, *Triumphant Democracy* (New York, 1886), p. 1.

4. Matthew Josephson, *Edison* (New York, 1959), p. 174. See also Thomas S. Kuhn, "The Relations Between History and History of Science," *Daedalus*, 100 (Spring 1971), 277, 283–84; and Oscar Handlin, "Science and Technology in Popular Culture," *Daedalus*, 94 (Winter 1965), 156–70.

5. Edward W. Byrn, *The Progress of Invention in the Nineteenth Century* (New York, 1900), pp. 4–6, 466.

6. George S. Morison, *The New Epoch as Developed by the Manufacture of Power* (1903; rpt. New York, 1972), esp. pp. 80–82.

7. Indeed one writer, David Hilton Wheeler, maintained in *Our Industrial Utopia and Its Unhappy Citizens* (1895) that late nineteenth-century America *was* utopia; only lack of "character" and desire for "superfluities" blinded people from recognizing the fact. Kenneth M. Roemer, "American Utopian Literature (1888–1900): An Annotated Bibliography," *American Literary Realism*, 4 (Summer 1971), 243.

8. Henry George, *Progress and Poverty*, 25th anniv. ed. (Garden City, N.Y., 1912), p. 8.

9. Norton to John Simon, Fall 1879, *Letters of Charles Eliot Norton*, ed. Sara Norton and M. A. DeWolfe Howe (Boston, 1913), II, 92; *Diary and Letters of Josephine Preston Peabody*, ed. Christina Hopkinson Baker (Boston, 1925), p. 73; Henry Adams, "A Letter to American Teachers of History," in *The Degradation of the Democratic Dogma* (New York, 1919), p. 195. See also John F. Kasson, "Medievalism in America during the Second Half of the Nineteenth Century" (B.A. honors essay, Harvard University, 1966).

10. Alvin Toffler, *Future Shock* (New York, 1970), p. 12.

11. Jay Martin, *Harvests of Change: American Literature, 1865–1914* (Englewood Cliffs, N.J., 1967), p. 142.

12. *The Education of Henry Adams* (Boston, 1918), p. 5.

13. Henry James, *The American Scene*, Indiana University Press Paperback (Bloomington, Ind., 1968), pp. 91, 156–57.

14. Lewis Mumford, *Technics and Civilization* (New York, 1934), p. 14.

15. *The Writings of Henry David Thoreau*, Walden ed. (Boston, 1906), II, 135, 130–31.

16. See Robert E. Riegel, "Standard Time in the United States," *American Historical Review*, 33 (October 1927), 84–89.

17. George M. Beard, *American Nervousness: Its Causes and Consequences* (New York, 1881), pp. 98–138.

18. See Roemer, "American Utopian Literature (1888–1900)," pp.

227–54; and Allyn B. Forbes, "The Literary Quest for Utopia, 1880–1900," *Social Forces*, 6 (December 1957), 179–80.

19. Howells's book appeared in serialized installments in *Cosmopolitan Magazine* from November 1892 through October 1893.

20. In discussing the writing of *Looking Backward*, Edward Bellamy observed that he never doubted that he would write the work as fiction: "This was not merely because that was a treatment which would command greater attention than others. In adventuring in any new and difficult field of speculation I believe that the student often cannot do better than to use the literary form of fiction. Nothing outside of the exact sciences has to be so logical as the thread of a story, if it is to be acceptable." "How I Wrote 'Looking Backward,'" in *Edward Bellamy Speaks Again!* (Chicago, 1937), pp. 223–24.

21. Northrop Frye, "Varieties of Literary Utopias," in *The Stubborn Structure: Essays on Criticism and Society* (Ithaca, 1970), p. 109. On fiction and the sense of apocalypse, see Frank Kermode, *The Sense of an Ending: Studies in the Theory of Fiction* (New York, 1967), pp. 3–31.

22. The phrase "the imagination of disaster" is Henry James's; Walter B. Rideout applies it to Ignatius Donnelly in his Introduction to Donnelly's *Caesar's Column: A Story of the Twentieth Century*, John Harvard Library (Cambridge, Mass., 1960), p. xxxi. See also Susan Sontag's essay, "The Imagination of Disaster," on science fiction films since World War II, in *Against Interpretation and Other Essays* (New York, 1966), pp. 209–25.

23. Bellamy, "How I Wrote 'Looking Backward,'" pp. 227–28; Lewis Mumford, *The Myth of the Machine: The Pentagon of Power* (New York, 1970), p. 218; Sylvia E. Bowman, *The Year 2000: A Critical Biography of Edward Bellamy* (New York, 1958), p. 121. On Bellamy's influence on other utopian writers in the late nineteenth century, see W. Arthur Boggs, "*Looking Backward* at the Utopian Novel, 1888–1900," New York Public Library *Bulletin*, 64 (June 1960), 329–36.

24. On the Nationalistic movement, see John L. Thomas's Introduction to Bellamy, *Looking Backward, 2000–1887*, John Harvard Library (Cambridge, Mass., 1967), pp. 69–85; Arthur E. Morgan, *Edward Bellamy* (New York, 1944), pp. 245–98; and Bowman, *The Year 2000*, pp. 119–38.

25. Bellamy, "How I Wrote 'Looking Backward,'" p. 218.

26. [Edward Bellamy], "Overworked Children in Our Mills," Springfield *Daily Union*, June 5, 1873, p. 2.

27. "With the Eyes Shut" was first published in *Harper's Monthly* in October 1889; it was reprinted in Edward Bellamy, *The Blindman's World and Other Stories* (Boston, 1898), pp. 335–65.

28. Thomas A. Edison, "The Phonograph and Its Future," *North American Review*, 126 (May–June 1878), 527–36.

29. John L. Thomas notes this symbolism in his Introduction to *Looking Backward*, p. 52.

30. Bellamy, *Looking Backward*, p. 141; *The Education of Henry Adams*, pp. 209, 238.

31. Morgan, *Edward Bellamy*, pp. 44, 145.

32. Bellamy, *Looking Backward*, p. 171.

33. Bellamy, *Looking Backward*, pp. 253, 304.

34. See Frederick Winslow Taylor, *The Principles of Scientific Management* (New York, 1911).

35. Bellamy, *Looking Backward*, pp. 177, 218.

36. Bellamy, *Looking Backward*, pp. 230, 212.

37. Bellamy, *Looking Backward*, p. 221.

38. Bellamy, *Looking Backward*, p. 115. In later statements Bellamy emphasized that his Nationalistic program would check excessive urbanization and enhance village as well as city life. Edward Bellamy, " 'Looking Backward' Again," *North American Review*, 150 (March 1890), 359. One should also note that *Looking Backward* inspired the Englishman Ebenezer Howard in the development of the Garden City. See F. J. Osborn's Preface to Howard, *Garden Cities of To-morrow*, M.I.T. Press Paperback (Cambridge, Mass., 1965), pp. 20–21.

39. Paul and Percival Goodman, *Communitas: Means of Livelihood and Ways of Life*, Vintage Books (New York, 1960), p. 125. See also Bradford Peck, *The World a Department Store: A Story of Life Under a Coöperative System* (Lewiston, Me., 1900), a fictional utopia influenced by Bellamy and written by the head of a large department store. For Peck's attempt actually to realize his utopia, see Wallace Evan Davies, "A Collectivist Experiment Down East: Bradford Peck and the Coöperative Association of America," *New England Quarterly*, 20 (December 1947), 471–91.

40. Bellamy, *Looking Backward*, p. 157.

41. Cf. Muzak's use of Mood Stimulus Curves in Kenneth Allsop, "Music by Muzak," *Encounter*, 28 (February 1967), 58–61.

42. Lewis Mumford, "Utopia, the City, and the Machine," *Daedalus*, 94 (Spring 1965), 277. See also Mumford's comments on *Looking Backward* in *The Pentagon of Power*, pp. 215–19.

43. See for example Richard Michaelis, *Looking Further Forward* (Chicago, 1890); Francis A. Walker, "Mr. Bellamy and the New Nationalist Party," *Atlantic Monthly*, 65 (February 1890), 248–62; and William Morris, *News from Nowhere* (London, 1891). Against charges of excessive militarism, Bellamy denied that the industrial army meant oppressive discipline, though he continued to defend and glorify the military analogy. Bellamy, " 'Looking Backward' Again," pp. 353–54, 357.

44. Bellamy to Howells, June 17, 1888, in Joseph Schiffman, "Mutual Indebtedness: Unpublished Letters of Edward Bellamy to William Dean Howells," Harvard Library *Bulletin*, 12 (1958), 370.

45. *Mark Twain–Howells Letters: The Correspondence of Samuel L. Clemens and William D. Howells, 1872–1910*, ed. Henry Nash Smith and William M. Gibson (Cambridge, Mass., 1960), II, 622n.

46. Sylvester Baxter, an enthusiastic supporter of Bellamy's Nationalism who accompanied him on his visit to Mark Twain's house, reviewed *A Connecticut Yankee* in the Boston Sunday *Herald* for December 15, 1889, and observed, "The record of [the Yankee's] adventures affords us another and very instructive sort of *Looking Backward*." Reprinted in Frederick Anderson, ed., *Mark Twain: The Critical Heritage* (New York, 1971), p. 151.

47. Dixon Wecter, *Sam Clemens of Hannibal* (Boston, 1952), p. 50.

48. Albert Bigelow Paine, *Mark Twain, A Biography* (New York, 1912), pp. 14, 725–28, 1056–58, 1098–99, 1150–51; George Hiram Brownell, "Mark Twain's Inventions," *The Twainian*, 3 (January 1944), 1–5.

49. Justin Kaplan, *Mr. Clemens and Mark Twain* (New York, 1966), p. 285; Bruce Bliven, Jr., *The Wonderful Writing Machine* (New York, 1954), pp. 58–62.

50. Mark Twain's involvement with the Paige typesetter is recounted most completely in Kaplan, *Mr. Clemens*, pp. 280–306. See also Paine, *Mark Twain*, pp. 903–14, 962–68.

51. Richard E. Huss, *The Development of Printers' Mechanical Typesetting Methods, 1822–1925* (Charlottesville, 1973), p. 78.

52. Mark Twain to Joe T. Goodman, June 22, 1890, in *Mark Twain's Letters*, ed. Albert Bigelow Paine (New York, 1917), II, 534.

53. Kaplan, *Mr. Clemens*, p. 284; Mark Twain to Howells, October 21, 1889, *Mark Twain–Howells Letters*, II, 615; Mark Twain to Joseph T. Goodman, October 7, 1889, *Mark Twain's Letters*, II, 516.

54. Mark Twain to Orion Clemens, January 5, 1889, *Mark Twain's Letters*, II, 506–508.

55. James M. Cox, *Mark Twain: The Fate of Humor* (Princeton, 1966), p. 209; Henry Nash Smith, *Mark Twain's Fable of Progress: Political and Economic Ideas in "A Connecticut Yankee"* (New Brunswick, N.J., 1964), pp. 59–60.

56. "Two Unused Prefaces to *A Connecticut Yankee*," Paine notebook 91, Mark Twain Papers, University of California, Berkeley, quoted in Roger B. Salomon, *Twain and the Image of History* (New Haven, 1961), p. 103.

57. Mark Twain, *A Connecticut Yankee in King Arthur's Court* (New York, 1889), p. 20.

58. Mark Twain, *A Connecticut Yankee*, pp. 95–96.

59. Mark Twain, *A Connecticut Yankee*, p. 85.

60. Jurgen Ruesch and Gregory Bateson, *Communication: The Social Matrix of Psychiatry* (New York, 1951), pp. 204n–205n.

61. A similar scene occurs in H. Rider Haggard's *King Solomon's Mines* (London, 1885), pp. 172–87, and may have suggested the idea to Mark Twain. Howard Baetzhold, who unaccountably overlooks this instance, cites as other possible sources Washington Irving's *Life of Columbus* (1829) and Emerson Bennett's popular novel, *The Prairie Flower* (1849). Baetzhold, *Mark Twain and John Bull: The British Connection* (Bloomington, Ind., 1970), pp. 346–47.

62. Mark Twain, *A Connecticut Yankee*, p. 212. See esp. John Ruskin's chapter, "The Nature of Gothic," in *The Stones of Venice*, II (London, 1853), 153–231. In "The New Dynasty," a paper Mark Twain delivered in 1886 while writing *A Connecticut Yankee*, he heralded the rise of industrially educated workers as sovereigns of a new age in history, with knowledge superior to kings and nobles of former times. The essay is reprinted in Paul J. Carter, Jr., "Mark Twain and the American Labor Movement," *New England Quarterly*, 30 (September 1957), 383–88.

63. On this point and the problems it raises see Harry B. Henderson

III, *Versions of the Past: The Historical Imagination in American Fiction* (New York, 1974), pp. 175–97.

64. Mark Twain, *A Connecticut Yankee*, pp. 120, 157, 160, 242.

65. Mark Twain, *A Connecticut Yankee*, pp. 97, 102, 395.

66. Mark Twain, *A Connecticut Yankee*, pp. 71, 95, 103. See Chadwick Hansen, "The Once and Future Boss: Mark Twain's Yankee," *Nineteenth-Century Fiction*, 28 (June 1973), 62–73.

67. Mark Twain, *A Connecticut Yankee*, p. 119.

68. Mark Twain, *A Connecticut Yankee*, p. 120.

69. For other volcanic metaphors in the book, see Allen Guttman, "Mark Twain's *Connecticut Yankee:* Affirmation of the Vernacular Tradition?" *New England Quarterly*, 33 (June 1960), 235. The sloppy layout and numerous manual typesetting errors in the excerpts of the newspaper the Yankee provides carried a special humor for Mark Twain, who was still eagerly supporting the Paige typesetter.

70. Thomas Blues, *Mark Twain and the Community* (Lexington, Ky., 1970), pp. 49–50.

71. Mark Twain, *A Connecticut Yankee*, pp. 355, 498.

72. Mark Twain, *A Connecticut Yankee*, p. 513.

73. Such analogies were underscored by Dan Beard's illustrations for the first edition, including one portraying Jay Gould as a slave driver with his boot on the throat of a manacled and prostrate woman; *A Connecticut Yankee*, p. 465.

74. Mark Twain, *A Connecticut Yankee*, p. 535.

75. Mark Twain read some of the relevant passages from *A Connecticut Yankee* to the West Point cadets on January 11, 1890. Mark Twain to Howells, December 23, 1889, *Mark Twain–Howells Letters*, II, 625.

76. The use of land mines in the Yankee's "Battle of the Sand Belt" recalls their similar use by other "Yankees" against the Confederates in the battle of Petersburg, Virginia, of July 1864. Ulysses S. Grant later described the "wild rumors" among the people of Petersburg during the siege in language similar to Hank Morgan's volcanic image of his industrial civilization: "They said that we had undermined the whole of Petersburg; that they were resting upon a slumbering volcano and did not know at what moment they might expect an eruption." Interestingly, Grant concluded that the effort was "a stupendous failure." *Personal Memoirs of U.S. Grant*, II (New York, 1886), 314. Mark Twain was undoubtedly familiar with this description, since he personally read Grant's manuscript and proofs and published the *Memoirs* through his publishing company in 1885–86, just as he was beginning *A Connecticut Yankee*. Kaplan, *Mr. Clemens*, p. 274.

77. Paul Wahl and Donald R. Toppel, *The Gatling Gun* (New York, 1965), pp. 12, 42. Mark Twain's report of his visit to the Colt arms works is reprinted in *The Twainian*, 7 (September–October 1948), 4. For examples of popular reports on the new technological warfare, see John Millis, "Electricity in Land Warfare," *Scribner's*, 6 (October 1889), 424–425; and William R. Hamilton, "American Machine Cannon and Dynamite Guns," *Century*, 36 (October 1888), 885.

78. See for example the discussion of firearms in Byrn, *Progress of Invention*, pp. 394–95.

79. Wahl and Toppel, *Gatling Gun*, pp. 80–82, 85.

80. Mark Twain, *A Connecticut Yankee*, pp. 178, 40.

81. Cf. the goal of an "electronic battlefield" in the Vietnam war and the problem of restraint it raised; Raphael Littauer and Norman Uphoff, eds., *The Air War in Indochina*, revised ed. (Boston, 1972), pp. 149–66.

82. Mark Twain, *A Connecticut Yankee*, pp. 542, 543. Cf. the London *Times* military correspondent's report on preparations for the Ashanti war in 1873 where he speaks of treating "a good mob of savages . . . to a little Gatling music"; reprinted in Wahl and Toppel, *Gatling Gun*, p. 67.

83. Mark Twain, *A Connecticut Yankee*, pp. 564, 554–56.

84. For my discussion of "psychic numbing" and unresolved guilt as responses to technological atrocity, I am indebted to the work of Robert Jay Lifton. See his essay "Beyond Atrocity," in *Crimes of War*, ed. Richard A. Falk, Gabriel Kolko, and Robert Jay Lifton (New York, 1971), pp. 15–27; *Death in Life: Survivors of Hiroshima* (New York, 1968); *History and Human Survival* (New York, 1970); and *Home from the War: Vietnam Veterans: Neither Victims nor Executioners* (New York, 1973).

85. Mark Twain, *A Connecticut Yankee*, p. 570.

86. On images of restoration, see Lifton, *Home from the War*, p. 367; and Lifton, "Images of Time," in *History and Human Survival*, pp. 58–80. On nostalgic elements in *A Connecticut Yankee* and their link with Mark Twain's recollections of his own childhood, see Henry Nash Smith, *Mark Twain: The Development of a Writer* (Cambridge, Mass., 1962), pp. 155–57.

87. Mark Twain, *A Connecticut Yankee*, p. 574.

88. *Mark Twain: The Critical Heritage*, pp. 149–50.

89. *Harper's Monthly*, 80 (January 1890), 320. Howells to Mark Twain, August 15, 1908, *Mark Twain–Howells Letters*, II, 833–34.

90. For different views of Mark Twain's attitudes toward *A Connecticut Yankee*, see James M. Cox, "*A Connecticut Yankee in King Arthur's Court*: The Machinery of Self-Preservation," *Yale Review*, 50 (Autumn 1960), 89–102; Charles S. Holmes, "*A Connecticut Yankee in King Arthur's Court*: Mark Twain's Fable of Uncertainty," *South Atlantic Quarterly*, 61 (Autumn 1962), 462–72; Smith, *Mark Twain's Fable of Progress*; and William K. Spofford, "Mark Twain's Connecticut Yankee: An Ignoramus Nevertheless," *Mark Twain Journal*, 15 (Summer 1970), 15–18.

91. D. H. Lawrence, *Studies in Classic American Literature*, Viking Compass (New York, 1964), p. 2.

92. On Donnelly's career, see Martin Ridge, *Ignatius Donnelly: The Portrait of a Politician* (Chicago, 1962).

93. On this subject, see Norman Pollack's interesting if at times strained argument in *The Populist Response to Industrial America: Midwest Populist Thought* (New York, 1962).

94. Ridge, *Ignatius Donnelly*, pp. 295–96.

95. Rideout, Introduction to Donnelly, *Caesar's Column*, p. xix; Ridge, *Ignatius Donnelly*, pp. 265–67.

96. For an extended comparison between *Caesar's Column* and *Looking Backward*, see Alexander Saxton, "*Caesar's Column*: The Dialogue of Utopia and Catastrophe," *American Quarterly*, 19 (Summer 1967), 224–238.

97. Donnelly, *Caesar's Column*, p. 61.

98. Donnelly, *Caesar's Column*, pp. 38, 39. The phrase "deathlike submission," which faithfully captures Donnelly's description, is from Jean-François Steiner's *Treblinka*; see Robert Lifton's review in *History and Human Survival*, pp. 195–207.

99. Donnelly, *Caesar's Column*, pp. 175, 173.

100. Donnelly, *Caesar's Column*, p. 259.

101. See John Patterson, "From Yeoman to Beast: Images of Blackness in *Caesar's Column*," *American Studies*, 12 (Fall 1971), 21–31.

102. Donnelly, *Caesar's Column*, p. 282.

103. Donnelly, *Caesar's Column*, pp. 245, 307.

104. H. L. Mencken, "The Dean," in *Prejudices, First Series* (London, 1921), pp. 52, 53.

105. See particularly Kenneth S. Lynn, *William Dean Howells: An American Life* (New York, 1971); and Lewis P. Simpson, "The Treason of William Dean Howells," in *The Man of Letters in New England and the South: Essays on the History of the Literary Vocation in America* (Baton Rouge, 1973), pp. 85–128.

106. Howells to James, October 10, 1888, in *Life in Letters of William Dean Howells*, ed. Mildred Howells (Garden City, N.Y., 1928), I, 417.

107. See Robert L. Hough, *The Quiet Rebel: William Dean Howells as Social Commentator* (Lincoln, 1959).

108. Howells to William Cooper Howells, February 2, 1890, *Life in Letters*, II, 1.

109. W. D. Howells, *The Altrurian Romances*, Selected ed. (Bloomington, Ind., 1968), p. 48.

110. Howells, *Altrurian Romances*, pp. 29, 62–63.

111. Howells, *Altrurian Romances*, p. 93.

112. Howells, *Altrurian Romances*, p. 147.

113. W. D. Howells, "Edward Bellamy," *Atlantic Monthly*, 82 (August 1898), 254.

114. Howells, *Altrurian Romances*, pp. 165–66.

115. Howells, *Altrurian Romances*, p. 158.

116. Cf. the values of Lewis Mumford's "neotechnic" civilization in *Technics and Civilization* (New York, 1934), pp. 364–435.

117. Howells, *Altrurian Romances*, p. 117.

118. Howells to Norton, April 15, 1907, *Life in Letters*, II, 242.

119. Smith, *Mark Twain's Fable of Progress*, pp. 40–41.

120. Edward Bellamy, Notebooks, Unpublished Papers of Edward Bellamy, Houghton Library, Harvard University, quoted in Thomas, Introduction to *Looking Backward*, p. 6.

Index

Page numbers in italics refer to illustrations.